INTRODU[...]

Robert A. Heinlein

"High Frontier" is the best news I have heard since VJ Day.

For endless unhappy years the United States has had no defense policy. We had something called a defense policy . . . but in the words of Abraham Lincoln, "Calling a tail a leg does not make it a leg."

Under our present policies what do we have? H-bombs, airborne, water-borne, and in silos, capable of destroying anything, anywhere on this planet. Elite troops second to none in our Marine Corps, in our Army's 82nd Airborne, and in our Navy SEALs. Other armed forces stationed around the world and on every ocean. Eyes in the sky that can spot any missile launched in our direction.

And none of these can even slow down an ICBM launched at Washington.

(Or at your hometown.)

So we have *no* defense. Instead we have something mislabeled a "defense policy," called "Mutually Assured Destruction," referred to as "MAD."

Never has Washington produced an acronym that fitted so perfectly. Picture two men at pointblank range each with a .45 aimed at the other man's bare chest. That is MAD. Crazy. Insane. And *stupid*.

High Frontier places a bulletproof vest on our bare

continued on page 7 . . .

HIGH FRONTIER

BY GENERAL DANIEL GRAHAM

A JIM BÆN PRESENTATION

TOR

A TOM DOHERTY ASSOCIATES BOOK

HIGH FRONTIER

Copyright ©1983 by High Frontier

A TOR Book

Published by:
Tom Doherty Associates, Inc.
8–10 West 36th Street
New York, New York 10018

First TOR printing, April 1983

ISBN: 523-48-078-4

Editorial Consultant: Dean Ing

Cover Design: Carol Russo

Printed in the United States of America

Distributed by:
Pinnacle Books
1430 Broadway
New York, New York 10018

TABLE OF CONTENTS

5

DEDICATION

This book is dedicated to President Ronald Reagan who, on 23 March 1983, made the historic decision to change the strategy of the United States from Mutual Assured Destruction to Assured Survival. This bold decision to deliver us all from the menacing balance-of terror doctrines of the past is ardently applauded by the scientists, engineers, strategists, and other specialists who are responsible for this book.

chest. High Frontier is as non-aggressive as a bullet-proof vest. There is no way to kill anyone with High Frontier—all that High Frontier can do is to keep others from killing us.

That is one of the two best aspects of High Frontier. It is so utterly peaceful that the most devout pacifist can support it with a clear conscience—indeed must support it once he understands it . . . as it tends to stop wars if war does happen. All who supported GROUND ZERO should support High Frontier.

The other best aspect of High Frontier is that its systems are nonnuclear. I am not one who gets upset at hearing the word "nuclear" . . . but no one in his right mind wants nuclear explosions going on over his head or anywhere on this planet. It is happy indeed that the best defense we can devise does not call for nuclear explosions. To save ourselves we do *not* need to blow up Moscow, we do *not* need to add to the radioactive fallout on our lovely planet.

The designers of High Frontier calculate that this new strategy will decrease our military costs. I am not in a position to judge this . . . but, frankly, I don't give a damn. A man with a burst appendix can't afford to dicker over the cost of surgery.

But will High Frontier in fact protect the Republic? As an old Research and Development engineer with the pessimism appropriate to the trade, I am certain that most of the hardware described in the High Frontier plan will undergo many changes before it is installed; that is the way R&D always works. But I am equally certain that the problems can be solved.

The first stages of High Frontier, point defense of our missile silos, we could start building later this afternoon; it involves nothing but well-known techniques and off-the-shelf hardware. That first stage alone could save us, as it denies to an enemy a free chance to destroy us by a preemptive first strike. It forces him to think twice, three times and decide not to try it.

But the key point is to whether this hardware will do the trick; the key lies in a change of attitude. A firm resolve to *defend* the United States . . . rather than resigning ourselves to the destruction of our beloved country. If we will so resolve, then the development of hardware is something we certainly know how to do.

But we won't get there by throwing up our hands and baring our necks to the executioner. God helps those who help themselves; he does not help those who won't try.

So let's try!
Robert A. Heinlein

HIGH FRONTIER

PREFACE

Jerry Pournelle

The Red Army has no recruiting posters. It doesn't need them. Not only are there no volunteers in the ranks of the Soviet Army, there can't be. There's no provision for volunteering for the ranks.

There's no need for volunteers, because every male Soviet citizen is conscripted at age 18. Every six months, approximately one million young men enter the system. They stay in for two years of training, after which they remain in the reserve registers until they reach the age of 50. They can be called up at any time. They number an estimated nine to eleven million.

There are always nearly two million men in the Land Forces alone; within ten days, these could be expanded to some 21 million. The Land Forces contain 180 divisions and 47 independent regiments of motor-rifle divisions. Each division has 23 tank companies and 67 artillery batteries.

The Soviet Strategic Rocket Forces muster some 325,000 troops. There are at least 1,400 land-based ICBM rockets. Five hundred intermediate range (IRBM) missiles are deployed in the Western USSR; this includes over 300 mobile SS-20s. The SS-20 can be reloaded, and many of the launchers already have at least one nuclear-tipped reload weapon.

In 1944, General Patton raced across France with his 3rd Army. The 16 divisions of First and Third Armies were supported by 5,600 trucks of the Red Ball Express. In 1975, the North Vietnamese Army moved south against Saigon. Its 20 divisions were supported by over 10,000 trucks and vehicles, nearly all of them sent into North Viet Nam from the Soviet Union. Those were the transport vehicles the Soviets could spare from their military establishment. The Red Army today has access to nearly half a million supply vehicles.

Germany entered World War II with 57 submarines. Britain had 58, Japan 56, and the United States 99. In 1941, the Soviet Union had 212 submarines in commission. They have about 275 submarines now, *in addition to* 83 Strategic Nuclear Forces nuclear subs. Of their 275 "regular navy" subs, at least 100 are nuclear powered.

We could continue, but surely the point is clear? The Soviet Union has built an enormous military machine, the largest peacetime military establishment in the history of mankind, and continues to maintain it. The expenses are great, but the Kremlin's control over the Soviet Empire is strong; apparently, the expense does not greatly concern the Politburo and its secret inner circle, the Defense Council.

We don't have to look into Soviet motives to conclude that the United States must respond to this enormous military buildup. The official policy of the Soviet Union is "world liberation". One may argue that they don't really mean it, and their revolutionary ardor long ago expired, but world revolution remains their official aim. If it is immoral to tempt a poor man by making theft easy, it seems no less so to tempt the Soviets by making conquest cheap and bloodless.

In fact, it is pointless to debate the issue. No responsible President or Congress can or will advocate leaving the United States defenseless in the face of the growing Soviet strategic threat. Unilateral disarmament may be a subject for debate within the population, but it has been overwhelmingly and repeatedly rejected by the American people, and our political leaders know this.

Granted that we must respond to the Soviet military threat, though, there remains the problem of what the response should be. It is no good responding ineffectively.

We could, if we chose, attempt to match the Soviets in men, machines, and weapons. Their military machine costs much less than ours, of course. As an example, they pay their soldiers no more than $25 US a month. Even so, the United States is far wealthier than the Soviet Union, and there is no question of our ability to *afford* a military establishment equal to or greater than theirs.

The costs would be high. Taxes would rise, and there would be real cuts in our standard of living; but we could do it. We could match the Soviets gun for gun, tank for tank, plane for plane, ship for ship.

We aren't likely to do that. Indeed, the events of the past two years demonstrate that the Congress isn't likely to do half that. Current administration efforts to bring the defense budget up to the proportions it held under John F. Kennedy have not been successful. Moreover, although the courts have held universal conscription to be constitutional, it certainly wouldn't be popular; and without universal conscription, we could never match the Soviets. We aren't *that* rich.

This, too, is a pointless debate; in the absence of some clear and unambiguous provocation, such as a massive invasion of Western Europe, or direct attack on Israel, the American people are not prepared to make the necessary sacrifices. We won't give up consumer goods, cosmetics, and the myriad luxuries we enjoy, nor will we opt for universal conscription. There is just no way that we'll respond to the Soviets by building a peacetime military establishment similar to theirs.

Unfortunately, although we have rejected matching the Soviet military establishment, we have not seized upon any viable alternative. Instead, we putter about, building some of this and some of that, hoping that our technological superiority will somehow do the trick even though we have no clearcut strategy of technology.

This has not always brought about good results. As Congressman Newt Gengrich, among others, has repeatedly pointed out, simply throwing money at the Pentagon is wasteful. Given money but no marching orders, the Pentagon almost always buys more M-1 tanks for the Army, more carriers for the Navy, wings of F-16s for the Air Force. They buy "things people can ride on," as one analyst recently put it.

Left to its own direction, the military is very conservative. Military establishments tend to keep the old, while flirting with the new and glamorous; to buy one or two armored cars, but keep horses for the cavalry. To put catapults and seaplanes on battleships, but reject aircraft carriers as not needed.

The result is a lack of direction. As the *Wall Street Journal* put it in November, 1982, "the Pentagon is an enormously inefficient nationalized industry. Its decisions are less the implementation of a coherent strategy than a matter of three services dividing a patronage pie. The most predictable result has been to deaden innovation."

The bold new systems are shunted aside; or, if the Pentagon is forced to deal with them, they are studied, tested, restudied, and retested. Then, suddenly, often as much a result of the geographical location of the factory that makes them as of strategic necessity, some of the most glamorous systems are procured.

We end up with weapons that no one is trained to use, aircraft with no spare parts and few trained pilots, communications systems that don't quite work, ships without trained sailors to man them, and missiles that work splendidly in test situations, but have profound problems on the battlefield.

I do not mean here to argue against high-technology weapons systems. One of the clearest lessons of the Viet Nam War was that high technology pays off. From "Blackbird" gunships to smart bombs and automatic mortars, high-technology weapons proved to have high effectiveness, and to be relatively cheap compared to the results they achieved.

The Falklands battles demonstrated the same point. High-technology weapons are essential for modern warfare. Moreover, the weapons must be in the hands of trained, able, and dedicated troops. It is not enough that we design and develop high-technology weapons systems. We must build them, deploy them, bring them to operational effectiveness, and maintain them. Anything short of that invites disaster.

However, it is not enough merely to recognize that high technology is vital to our future. There must also be a strategic focus. As Stefan Possony and I have argued elsewhere, we must have strategic direction to our military research and development. We must have a strategy of technology.

That won't be developed overnight. Most analysts believe it will require a grueling and painful reorganization of the entire defense establishment. That will generate great opposition. There are too many vested interests for things to be otherwise.

It will also require time. We may not have that time. Military establishments, ours among them, have *always* been inefficient, and better organized for the last war than the next. If we wait for perfection, we may well wait forever.

Thus three facts stand out:

1. The Soviets have an enormous military establishment, and we are not going to match it tank for tank and gun for gun.

2. Our present course of buying some of this and some of that, more tanks here and more planes there, isn't an adequate, or indeed reasonable, response to the threat, and "reform of the Pentagon" and other efforts to "trim the fat and reduce waste" aren't likely to succeed very quickly, if at all.

3. We have to do *something*, and soon.

This reasoning was the starting point for Lt. General Daniel O. Graham's strategic analysis. If what we're doing isn't going to work, and we have to do something, where can we go? Graham concluded that we needed a bold new approach, a strategic sidestep; that we had to stop competing with the Soviets in areas in which we can't win, and begin to compete where we have the advantage.

His analysis led him through high technology to space; to the High Frontier. As Dan Graham has repeatedly said, he didn't start with any prejudices toward space as a decisive frontier. All his training and experience pointed him elsewhere. It was the search for strategic initiatives which led him to his conclusions.

This book presents, in detail, a bold new strategy for the defense of the United States and Western Civilization. The concepts of High Frontier are very new and different, but the plans suggested here are realistic. Some of the nation's best engineers and development scientists have examined Project High Frontier. Many began their analysis convinced that High Frontier couldn't work, or would cost too much, or would take too long. As they became more involved, they changed their minds.

I know, because I was one of them. I am not any longer a professional scientist, but I stay in touch with the aerospace community. When I was first told of High Frontier, I searched among those I respected for engineers and scientists opposed to the plan, and introduced them to the team of equally respectable advisors assisting General Graham. In some cases I was privileged to sit in on the resulting debates.

Certain conclusions emerged. First, we do hold the advantages in space. Our technology is more advanced and more reliable, and the Soviet "brute force" approach to problems creates as many difficulties as it solves when applied to the space environment. We don't have everything our way in space, but we are clearly better off in a high-technology competition than trying to match their conventional military establishment.

Secondly, High Frontier will work. There can be arguments over details, costs, and schedules. As with all strategic plans there will remain uncertainties: the first thing taught to career officers is the maxim, "No battle plan ever survives contact with the enemy." But it will work, and those who all too predictably argue that High Frontier's supporters don't understand the laws of physics are cordially invited to present their case, not to Dan Graham, but to the prizewinning physicists aboard General Graham's team.

In *The Strategy of Technology*, Stefan Possony and I argued that the United States ought to abandon the doctrine of Mutual Assured Destruction, sometimes known as MAD, in favor of a strategic doctrine of "Assured Survival"; that as a Western nation adhering to the Judeo-Christian tradition, we should be more concerned with preserving our nation than with assuring another's destruction.

Project High Frontier presents a practical way to achieve that goal.

J. E. Pournelle, Ph.D.

Dr. Pournelle was for over fifteen years a professional scientist and systems analyst within the aerospace community. He was general editor of the Top Secret study *Project 75*, the most comprehensive study of strategic doctrine and missile technology then in existence, and principal investigator of "Stability and Strategic Doctrine", a study conducted for Headquarters, United States Air Force, and presented to the Air Council. His book (written with Stefan T. Possony) *The Strategy of Technology* has been used as a text in the US Air Force Academy and the Air War College.

Dr. Pournelle is a Fellow of the Royal Astronomical Society and of the British Interplanetary Society, and a Member of the London-based Institute for Strategic Studies. He is Chairman of the Citizens' Advisory Council on National Space Policy.

FOREWORD
Lt. Gen. Daniel O. Graham, USA (Ret)

Americans of all political stripes and walks of life have every right to perceive the thundering debates in Washington over ways to deploy the MX missile and percentages of GNP necessary for defense as symptomatic of some sort of mental aberration—a sign of madness. A commonsensical person can with all probity ask just what that x or y percent of the national budget or 20 billion dollars' worth of concrete poured around MX missile silos can do to defend *him* from the destruction of war. When he does pose such questions, he is apt to be surprised by utterly contradictory opinions from real and phony "experts" whose *actual* motives are not to inform him but to drive him into one or another political camp. His choice seems either to trust the experts from Government who "ought to know what they're doing" or to trust their vociferous opponents who seem to oppose *every* defense program posed.

The average American is right when he balks at giving his full trust either to the bureaucracy or to the pacifists. He is also right when he considers the clash of these views in Washington as a form of madness, for MAD—Mutual Assured Destruction theory—lies at the root of most of it.

High Frontier prescribes the cure for the madness.

21

High Frontier is a privately funded effort. Its purpose is to seek answers in U.S. technology, especially space technology, to the strategic problems that plague the United States and the Free World.

The origins of this effort lie back in the days when I was a military advisor to then-candidate Ronald Reagan. Early in the campaign I was among those insisting that the only viable approach for a new administration, when coping with the growing military imbalances, was to implement a basic change in the U.S. grand strategy—to make a "technological end-run on the Soviets."

As far as I could determine, all advisors to Mr. Reagan agreed with this conclusion—at least in principle—at the time. However, as time passed, this more fundamental approach to national security issues receded into the realm of theory. The team of advisors on security matters began to concentrate instead on the amounts of money needed to revitalize ailing, ongoing Pentagon programs, and on the "quick fixes" necessary if the U.S. was to hold its own within the context of *current* strategy and doctrine. The Carter Administration's defense budget was checked, line item by line item, with a view to repairing past damage to U.S. capabilities. We recommended new expenditures to plug, as quickly as possible, the strategic gaps between U.S. and Soviet capabilities which are known collectively as "the window of vulnerability."

Some of the team continued to believe that a strategically and technically sound alternative to this incremental approach could be found. None of us could be sure, at the time, what that alternative might be.

Then in early 1981, Congressman Newt Gingrich of Georgia and I discussed the future of the new Administration in the area of national security. Mr. Gingrich shared my apprehension that large Department of Defense budget increases would not, in themselves, solve our military problems—and, in fact, that the budget increases might not be sustained even by the new pro-defense Congress for more than two years. We discussed the possibilities of setting forth a new strategic approach and a technological end-run on the Soviets, to meet President Reagan's commitment to a margin of safety. We threw our energies into the formulation of this new approach.

The fundamental strategy change required was in replacement of the Mutual Assured Destruction (MAD) doctrine which had shaped—rather, *warped*—our strategic force posture, and had undergirded the U.S. approach to arms control. The MAD doctrine postulates that strategic *defensive* systems are destabilizing and provocative—a theory that has led to serious vulnerabilities in the Free World to nuclear attack and blackmail.

Although military spokesmen had from time to time denied that MAD was U.S. policy, our political machinery adhered to it in essence. Certainly SALT negotiations had been conducted as if MAD were official U.S. strategy. SALT I attempted to disallow strategic defense by negating any significant antiballistic missile efforts. On the other hand, SALT I accommodated a massive offensive nuclear buildup then underway in the USSR, and permitted a proliferation of city-busting MIRVs on our side.

U.S. negotiators later accepted a SALT II treaty which was badly flawed in detail and which, in general, merely expanded the limits on offensive nuclear power to accommodate obvious Soviet programs. Before our

Senate rebelled and permitted SALT II to die, the standard rebuttal of Carter Administration spokesmen to critics of the treaty was a dictum based on MAD: we would still be able to kill millions of Soviet citizens in a retaliatory strike with weapons we already had, and could add more if we wished to do so under SALT II.

A search for technology which would provide the basis for an end-run on the Soviets led inexorably to space. The U.S. advantage in space is demonstrated in its most dramatic form by our Space Shuttle. More fundamentally, the ability of the U.S. to miniaturize components gives us great advantages in space, where transport costs-per-pound are critical. Today, a pound of U.S. space machinery can do much more than a pound of Soviet space machinery.

Fortunately, the technologies immediately available for military systems in space—beyond intelligence, communication, and navigation-aid satellites—are primarily and directly applicable to ballistic missile *defense* systems. This fact strongly implied that our space programs held the key to a technological end-run which would offset current Soviet strategic nuclear advantages and, at the same time, could provide an escape from the MAD balance-of-terror doctrine.

Early in 1981, I wrote an article titled, "Toward a New U.S. Strategy: Bold Strokes Rather than Increments," which was published in the Spring issue of *Strategic Review*. This article laid out the basic concept of a spaceborne defense which would nullify the MAD doctrine.

Although I was convinced that spaceborne defenses are feasible (perhaps using lasers and other beam weapon technology), I was unable to conceptualize a beam weapon system which could withstand close scrutiny in terms of reasonably short lead time. However, in consultation with conceptual and technical experts working on other military space applications, we came up with a concept for a spaceborne ballistic missile defense system which entails two great advantages: it is based largely on developed technology; and it is nonnuclear in character.

In order to avoid long lead times and interminable arguments among scientists, we sought to use off-the-shelf technology as much as possible. And in order to avoid fruitless searches for perfection, we postulated that the system would suffice if it put at risk as little as 20 % of an all-out missile attack on the U.S. We used this postulate because even that modest level of attrition of a Soviet missile attack *in the early stages of trajectory* would be sufficient to destroy any confidence Moscow might have in a disarming first strike.

The solution we found was a spaceborne missile defense concept which can put at risk a much higher percentage of a Soviet strategic missile salvo, whether that salvo is fired against the U.S. or our allies. The system concept, using many off-the-shelf components, appears comparatively inexpensive and can probably be deployed in a relatively short time. This system is especially attractive in view of the fact that we need not wait for the whole system's development, to deploy some of its elements effectively.

Thus far, the global ballistic missile defense system concept we call the High Frontier has held up well under severe scrutiny for feasibility, costs, timing, and vulnerability. *It may or may not be the best technical option available to us.* At a minimum, it has demonstrated the basic feasibility of spaceborne defenses which can fundamentally change the nature of the strategic balance from Mutually Assured Destruction toward Assured Survival.

In the search for military options in space we were fortunate to rely on Brigadier General Robert Richardson (USAF Retired); the Honorable John Morse, former Deputy Assistant Secretary of Defense; and Dr. Arnold Kramish, our scientific advisor. Further, we are indebted to a group of Boeing Company engineers who rechecked our findings and provided invaluable advice.

Our horizons were expanded by Dr. Peter Glaser of Arthur D. Little Company, who convinced us that space held the key, not only to national security but also to economic growth and energy supply as well. As a result of his input to our efforts, the High Frontier concept was broadened to constitute a true national strategy, rather than a purely *military* strategy. We came to realize that the U.S. could best realize its opportunities in space if military and civil space developments proceeded together, yielding very large economic returns for our initial investments. Indeed, the attempt to separate these elements in government, and in public groups supporting space efforts, was a prime reason for the lack of a vigorous, purposeful U.S. space effort. The Reagan Administration has taken laudable steps toward correcting this conceptual flaw.

In the Fall of 1981, High Frontier became a project of the Heritage Foundation where it has profited from the strong support of Edwin Feulner, Jr., President.

Frank Barnett gave a great boost to the High Frontier concept by providing us the opportunity to present it to an audience of distinguished citizens assembled in Washington by his National Strategy Information

Center. The Honorable Karl R. Bendetsen was in the audience and spoke to me of his enthusiasm for the High Frontier. He contributed substantially to the on-going momentum of the project, and to its definition and consensus.

This book was originally compiled by the persons listed below. Each contributed to the effort not only in a given area of expertise, but also to the development of the basic High Frontier concept.

Dr. Jeffrey Barlow	The Heritage Foundation Liaison
John Bosma	Space Doctrines
John J. Coakley	Public Relations
Dr. Miles Costick	Technology Transfer
Prof. James Dougherty	International Relations
Dr. Jacqueline K. Davis	Cruise Missiles
Col. Sam Dickens, USAF (Ret.)	Affected Treaties
William J. Gill	Public Information
Dr. Peter Glaser	Nonmilitary Space Systems
Lawrence Hafstad	Industry
Dr. Mose Harvey	International Reactions
Frank Hoeber	Military Costs
Orlando Johnson	Macroeconomics
Cresson Kearny	Civil Defense
Brig. Gen. Albion Knight, USA (Ret.)	Public Reaction
Arnold Kramish	Defense Technology
Cleveland Lane	Communications
Sherri Mayerhofer	Administrative Chief
Vicki McCowan	Administrative Ass't., Editing
Marianne Mele	Legal Affairs and Editing

Maj. Gen. Stewart Meyer, USA (Ret.)	Ballistic Missile Defense
Ed Milauckas	Implementation
Hon. John Morse	Collateral Systems
Dr. Ralph Nansen	Space Systems, General
Dr. Robert Pfaltzgraff, Jr.	Alliance Reactions
Dr. Jerry Pournelle	Organizational Support
John Rather	Advanced Systems
Fred W. Redding, Jr.	Military Space Systems
Brig. Gen. Robert Richardson, III, USAF (Ret.)	Implementation
Dr. Peter Vajk	Space Economics
James Wilson	Advanced Technology Systems

It is most gratifying that these people served—and still serve—High Frontier for no compensation, or less compensation than they deserve or could receive elsewhere. This expression of concern for U.S. national interests and of support for the High Frontier concept is heartening. We are further indebted to the generosity of individuals and institutions whose major contributions made the High Frontier Study possible.

This volume is the result of High Frontier thus far. There is much more to do before the promise becomes reality.

Daniel O. Graham
Washington, D.C.
April 1983

CHAPTER I
AN OVERVIEW OF
THE HIGH FRONTIER SYSTEM

The United States has an historic, but fleeting, opportunity to take its destiny into its own hands. We can reverse some ominous military and economic trends which today beset the peoples of the Free World and can restore confidence in the future of free political and economic systems.

To accomplish this, we need only make maximum use of one priceless advantage—superiority in space technology. We can escape the brooding menace of offensive "balance of terror" doctrines by deploying nonnuclear defensive systems in space. We can further confound the prophets of doom by opening the vast, rich High Frontier of space for nonmilitary industrialization. Beginning with a vitally necessary shift in national strategy, we can achieve tremendous gains for mankind through peaceful development of nonpolluting energy sources for all to share.

If we are to seize this historic opportunity, we must first muster the political will to discard without qualm the failed doctrines of the past, to aggressively overcome bureaucratic impediments to action, and to meet, without flinching, the wave of indignation from outraged ideologues at home and abroad. The basic technology is available and the costs are reasonable. We see no

other promising alternatives to our security problems
—certainly none with such enormous potential for
global benefits in a peaceful context.

GOALS AND GUIDELINES

The objective of the High Frontier system is to
apply a national strategy option which would make
maximum use of U.S. space technology to accomplish
the following goals:

- Nullify the growing threat to the U.S. and its
  allies which is posed by Soviet offensive military
  power.

- Replace the dangerous offense-oriented doctrine
  of Mutual Assured Destruction (MAD) with a
  defensive strategy of Assured Survival, which maximizes avoidance of war by greatly reducing an
  aggressor's confidence in his first-strike capability.

- Provide both security and incentive for realizing
  the enormous industrial and commercial potential
  of space.

In conceptualizing a system to meet these goals,
we maintained primary guidelines, i.e., that every
facet of the system must be militarily sound, technologically and economically feasible, and politically
practical. We believe that these guidelines were scru-
pulously maintained—though we fully expect to encounter a great deal of resistance in the political
arena.

THE FUNDAMENTAL THREAT

The High Frontier studies focused primarily on countering the Soviet military threat which is ominous and growing. This threat is the result of determined efforts by the Soviet Union to establish global military dominance—efforts that have been abetted by poorly conceived U.S. security policies such as MAD. The Soviet military buildup coupled with U.S. military neglect has created these alarming conditions:

- There is a serious and growing Soviet advantage in strategic nuclear weapons which cannot be countered by the undefended U.S. except by a threat of retaliation that involves national suicide.

- The preponderance of Soviet conventional weapons vis-a-vis the U.S. and its allies is also severe and growing. It can no longer be counterbalanced, as it has been in the past, by a credible threat to bring higher technology U.S. weaponry to bear.

- The Soviet Union is increasingly successful in the use of propaganda and the application of direct or indirect military power to disrupt our alliances, and to force Marxism on underdeveloped countries. This Soviet success now threatens the continuing availability of raw materials which are critical to the industrialized West.

- The West is dangerously dependent on diminishing oil supplies located in areas threatened by Soviet military or manipulative political power.

- The U.S. alliance system is in serious disarray. It suffers a lost sense of purpose and a perception of a decline in U.S. power and leadership. The Soviet propaganda offensive against U.S. nuclear weapons was designed to persuade our European allies to become neutral, and that offensive is increasingly effective.

The Soviets are engaged in a costly and all-too-successful effort to cap their current strategic advantages —in their terms, "a favorable correlation of forces" —with Soviet domination of near-Earth space. The Soviets have the only tested space weapon on either side, an antisatellite system. They have orbited nuclear reactors. They have a manned space station in orbit and are expanding it. Indeed, as of early 1983, the Soviets are hinting at a spectacular new space development which may amount to a Soviet battlecraft in near-Earth orbit. Almost all Soviet space activity has a distinct military flavor. The essence of the Soviet military space threat was included in the 1981 Department of Defense publication *Soviet Military Power* (pp. 79–80):

The Soviets have a vigorous and constantly expanding military space program. In the past ten years they have been launching spacecraft at over 75 per year, at the rate of two-to-five times that of the United States. The annual payload weight placed into orbit by the Soviets is even more impressive—660,000 pounds—ten times that of the United States. Some, but by no means all, of the differential can be accounted for by long-life U.S. satellites using miniaturized high-technology components. Such an activity rate is expensive to underwrite, yet the Soviets are willing to expend resources on space hardware at an approximate eight % per year growth rate in constant dollars. We estimate that 70 % of Soviet space systems serve a purely military role, another 15 % serve dual military/civil roles, and the remaining 15 % are purely civil. The Soviet military satellites perform a wide variety of reconnaissance and collection missions. Military R&D experiments are performed onboard Soviet manned space stations, and the Soviets continue to develop and test an ASAT antisatellite co-orbital interceptor. The Soviets appear to be interested in and possibly developing an improved ASAT. A very large space booster similar in performmance to the Apollo program's Saturn V is under development and will have the capability to launch very heavy payloads into orbit, including even larger and more capable laser weapons. This booster is estimated to have six-to-seven times the launch weight capability of the Space Shuttle. Soviet space research and development, test, production, and launch facilities are undergoing a continuing buildup. The new booster will be capable of putting very large permanently manned space stations into orbit. The Soviet goal of having continuously manned space stations may support both defensive and offensive weapons in space with men in the space stations for target selection, repairs and adjustments and positive command and control. The Soviet's predominantly military space program is expected to continue to produce steady gains in reliability, sophistication and operational capability.

The Soviets consider space a perfect environment in which to exercise their long-standing doctrinal and operational preferences in warfighting—unconventional "first moves," preemptive attacks or "decapitation attacks" against vital targets such as strategic communications, "combined-arms" moves (made possible by ship-tracking satellites), and other elements of their well-stocked repertoire. The Soviets integrate military space operations into their strategic thinking. They see space in straightforward terms, as an operational or combatant theater, whereas we see it—given our own strategic culture—as a "sanctuary" where support forces for terrestrial military forces can operate without fear of direct attack.

If Moscow achieves its aims, we will be faced with a new era of *Pax Sovietica* in which Soviet space power dictates Free World behavior. We believe that the High Frontier of space provides us with the opportunity— perhaps our only opportunity—to frustrate Soviet power ambitions and at the same time open up a new era of hope and prosperity for the U.S. and the Free World.

The immediate threat impels us to exploit our space technology, but we also have an unavoidable historical imperative to move vigorously into that arena. Throughout man's history, those nations which moved most effectively from one arena of human activity to the next have reaped enormous advantages. For example, when man's activities moved from land to the coastal seas, the Vikings established an extraordinary dominance by excelling at sailing those seas.

After the epic voyages of Columbus and Magellan, Spain and Portugal dominated the world through military and commercial control of the new arena of human activity—the high seas. Later, England with her powerful fleet of merchantmen and men-of-war established a century of *Pax Britannica*. When the coastal seas of space—the air—became a new sphere of human activity, the United States gained great strategic advantages by acquiring the most effective military and civilian capability in aviation. Today, after epic manned and unmanned explorations of space, we shall see which nation puts the equivalent of the British merchantmen and men-of-war into space. We dare not let it be our adversary.

HIGH FRONTIER'S MILITARY DIMENSION

We cannot reverse the ominous trends in the military balance if we adhere to current strategy, trying to compete with the Soviets in piling up weapons of current technology. Even if Congress were willing to appropriate unlimited funds to procure these weapons (and it is *not*), our defense production base is in such a sorry state that it could not compete with the current high-level output of the Soviet arms-production base. Our best hope is to change our strategy, and to move the key competition into a technological arena where we have an advantage.

A bold and rapid entry into space, if announced and initiated now, would end-run the Soviets in the eyes of the world and would move the contest into an arena where we could exploit the technological advantages we still hold. This is far preferable to pursuing an offensive-weaponry numbers contest here on Earth—a contest which would be difficult if not impossible for us to win. We are suggesting a different strategic contest, defensive and nonnuclear, which we *can* win.

When we look to space for the technological end-run on the Soviets, we find that all factors call for emphasis on strategic defense. First, defensive systems hold the only promise to break out of the MAD doctrine. Second, defense is the only sound alternative to costly "Racetrack" or "Dense Pack"-type options to protect our deterrent systems. Third, our current and crucial military investment in space is also vulnerable to attack. Fourth, available technology favors defensive space systems. Fifth and last, there are severe political constraints and some technical-military reasons inhibiting the deployment of offensive weapons—especially nuclear weapons—in space.

For these reasons, the military dimension of High Frontier emphasizes the resurrection of a long-neglected aspect of our security: protective strategic defense. We visualize a layered strategic defense. The first layer would be a spaceborne defense which would effectively filter an antagonist's missile attack in the early stages of trajectory. The second layer would be a broader space protection system, perhaps eventually using advanced beam weaponry to further reduce the effectiveness of a missile attack and to defend other space assets from a variety of attacks. The third layer would be a groundbased point defense system capable of removing any antagonist's assurance of a first-strike success against our missile silos. This final layer of defense could prove an effective deterrent even *before* the first and second layers of our system are deployed. In the completed system, as backup to those spaceborne layers, the groundbased point defense layer would intercept hostile missiles which might leak through the spaceborne defenses. A passive fourth layer would be civil defense, which becomes a valuable aspect of strategy in conjunction with these three active defense layers. *None of these layers of defense need employ nuclear warheads.*

We can get a point defense layer within two or three years which would be adequate to protect our ICBM silos and which can avoid the high-cost deployment modes for MX. An initial spaceborne global ballistic missile defense (GBMD) can probably be acquired in five or six years, given adequate priority. A second-generation general space defense using more advanced technology can probably be achieved in the early 1990s.

In proposing such strategic defenses, one invariably encounters the shibboleths that have plagued defensive strategies in the past. It has been an article of faith in the offense-only, Assured Destruction school of thought that strategic defenses in the nuclear era are useless unless they are impermeable or not subject to attack, and/or that they are impossibly expensive. These are false premises, as we will show.

One issue which must be carefully addressed is that of space-system survivability. While space systems are nearly invulnerable to a large array of threats with which terrestrial systems must cope (e.g., bombs and inclement weather), they have some unique vulnerabilities to threats which can be posed by a technologically advanced adversary. An examination of this problem leads to several conclusions:

- As with other systems, no spaceborne system can be envisaged which is invulnerable to *all* postulated threats.

- Vulnerability of current U.S. space assets (intelligence and communications satellites and the

Shuttle) sharply increases the need for an effective spaceborne defensive system which can defend itself and reduce the threat to other space systems, as well as defend ground targets against hostile objects transiting space, e.g., ICBMs.

- Defensive systems employing large numbers of less sophisticated satellites are far less vulnerable than those employing smaller numbers of more sophisticated satellites.

- Very important to survivability is the ability to provide mutual warning and protection among satellites in a ballistic missile defense.

- The sooner a spaceborne defense system can be deployed against ballistic missiles, the better its survivability. Long lead-time systems are susceptible to long lead-time Soviet countermeasures, real or postulated.

- Future U.S. deployment of more sophisticated beam weapon military satellites may depend, for survivability, on protection provided by a lower-technology defensive system already deployed.

Given the characteristics of currently operating U.S. space systems, one can readily imagine ways for the Soviets to attack them, ranging from throwing sand in their paths to incinerating them with futuristic beam weapons. Such attack modes fall into two basic categories: peacetime attack and wartime attack.

Most current Soviet capabilities to attack U.S. space systems are applicable in the peacetime attack category. These include attack with nonnuclear direct-ascent missiles, the current Soviet antisatellite system, and current power-level Soviet lasers. However, these attack modes presuppose Soviet willingness to risk the grave consequences (including war) of attacking our space systems in time of peace or crisis. While such Soviet action cannot be totally ignored, most experts on Soviet behavior find this possibility extremely remote.

The second class of threat, wartime attack, is more serious. In this situation, nuclear weapons could be used to destroy or disable our space systems using radiation effects, including electromagnetic pulse effects. (Blast effects are of little consequence outside the atmosphere.) There are technical means for reducing the vulnerability of space systems to these effects, but the best counter to such threats is the capability of a defensive system to intercept hostile objects directed at it.

The Soviets may develop laser weapons of such power that satellites passing over them could be destroyed with a single burst of energy. It is doubtful, however, that such systems could, in the foreseeable future, successfully attack satellites coming over the horizon where they would be shielded by much more of the Earth's atmosphere. Laser beams are severely attenuated in passing through many miles of atmosphere.

Probably the most important factors in the survivability problem are military rather than technical. Survivability is sharply increased by the ability of space vehicles to destroy threatening objects launched at them, or at other U.S. space vehicles. Even if the Soviets eventually create the means to attack a spaceborne defense system in order to launch a subsequent strategic missile attack, they would lose all chance of destroying the U.S. groundbased deterrents. In those circumstances, launch-on-warning or launch-under-attack become both credible and feasible counterattack options for the U.S. The Soviets could not expect, after attacking us in space, that the U.S. President would hesitate to respond to sensor warnings that a missile attack had been launched from the USSR. This fact alone would make a spaceborne defense of great strategic value, greatly strengthening the deterrent to nuclear war.

The urgent requirements for military systems to implement the High Frontier system are these:

1. A point defense for U.S. ICBM silos which, within two or three years, and at a cost less than that of superhardening, can destroy any confidence the Soviets might have in a first strike against our deterrent.

2. A first-generation spaceborne ballistic missile defense, deployable in five or six years at a cost not exceeding that of previously proposed MX missile deployment schemes, and capable of serious attrition of a Soviet strategic missile attack in the early part of trajectory.

3. A second-generation space defense system, deployable within 10 or 12 years and capable of attacking hostile objects anywhere in near-Earth space with advanced weaponry.

4. A utilitarian manned military space control vehicle, deployable within the next six to eight years, and capable of inspection, on-orbit maintenance, and space tug missions wherever satellites can go.

5. A civil defense program of sufficient scope and funding to take advantage of the proposed active strategic defenses, and thus to add to U.S. deterrent strength.

HIGH FRONTIER'S NONMILITARY DIMENSION

Space holds the promise of a new era in economic expansion. The unique environment of space—zero gravity, near-perfect vacuum, unlimited heat absorption, sterile conditions—opens up a broad range of industrial/commercial possibilities. Space also contains virtually inexhaustible supplies of minerals and solar energy. The economic potential of space is already being tapped in the communications industry. As the cost of space transportation is lowered, the industrialization of space can expand tremendously. However, the capital investment in space industries must be quite large, and such investments are unlikely if space installations are indefensible against attack. For this reason, military capabilities in space are crucial to spacebased economic growth.

We should harbor no illusions that space can be limited to "peaceful uses" any more than could previous arenas on land, sea, or in the air. Indeed, most current space assets, U.S. and Soviet alike, are partially or entirely military—and the most destructive weapons of all, nuclear-tipped ballistic missiles, must transit space en route to their targets.

The government's role in opening up the High Frontier of space for economic exploitation is basically the same as it has always been with the opening of frontiers: exploration, transportation systems, and security. These functions translate into certain specifics: scientific research, improving the Space Shuttle, and providing spaceborne defenses.

Both the military and nonmilitary uses of space depend on continued efforts in core technologies: improvements in space transportation to reduce the cost-per-pound of materials in orbit, and the creation of permanent, manned space stations at the "terminals" of the space transport system.

While these efforts are initially the responsibility of government, they should be undertaken in cooperation with private industry and with support from other benefiting nations. With a proper combination of space technologies, we can sharply improve the security of the U.S. and its Free World allies; and, at the same time, restore confidence in the ability of Free World economies to meet challenges of the future.

The urgency here is far greater than many people in this country appear to realize. Following the success of the U.S. moon landing, the Soviets made it clear that, while intending first and foremost to develop maximum possible military capabilities in space, they expected also to achieve dominance in economic exploitation of space. In 1964 Brezhnev spoke of these plans, and Soviet specialized literature has gone into great detail concerning concrete possibilities. Further, all phases of ongoing Soviet space activities that aim at strategic objectives also serve as stepping-stones to the USSR's preeminence in space for nonmilitary as well as military purposes.

The primary urgent nonmilitary requirements for the High Frontier system are these:

1. Improved space transport, designed to lower the cost-per-pound in orbit to under $100.

2. A manned space station in low-Earth orbit as soon as practicable. This would allow low-cost, efficient development and testing of both civilian and military system elements, and would constitute a

first step toward a similar manned station in geo-synchronous orbit.

3. Development work on reliable, high-capacity energy systems in space. Initially this energy system would power other space activities (as opposed to a nuclear reactor, for example). Eventually it would provide electrical power from orbit to any spot on Earth. Solar Power Satellites (SPS) promise a new abundance of energy for all nations; power which is nonpolluting and, in the long run, very inexpensive.

4. Preparatory development of a selected number of promising commercial business opportunities. Government efforts should focus on encouraging the transformation of the "seed" efforts into independently viable commercial operations as soon as possible.

All of these requirements can be met; some with technology already in hand and off-the-shelf hardware. None of these requirements demand technological breakthroughs or any commitment to mere scientific theories. Our optimism with regard to the system is, of course, a consensus. A few analysts of the High Frontier system have expressed concern that some developments may take longer than currently projected, e.g., transmission rates for encrypted information for control and coordination of space systems might not initially be as high as deemed optimum. But even here the debate is more a matter of optimization than of feasibility. Both in the military and nonmilitary arenas of fast data transmission, we are steadily progressing. The serious caveats deal not with *if*, but with *when*— further grounds for optimism when dealing with a projected new system.

There are, in fact, a variety of viable options available to meet each major requirement of the High Frontier. In the following pages we will present a description of one set of programs which would meet the requirements. Each element is further described in later chapters of this book. The costs estimated for these programs are in constant dollars. Costs and times indicated are based on a management system which minimizes bureaucratic delays.

PROGRAM ELEMENTS OF THE HIGH FRONTIER

Clearly, the development programs will vary in complexity and lead-time. Yet it is worth repeating that we can deploy most of the system elements as they come "on-line," and some of these elements are already under serious study. These programs include a point defense layer; spaceborne defense layers; temporarily increased civil defense funding; a multipurpose spaceplane; an upgraded version of our current space shuttle; a manned space station; and a pilot version of an SPS for continuous electrical power.

Perhaps the most quickly deployable element of the system is the point defense layer which would be emplaced near our own deterrent missile silos. A partially tested system exists that could meet the requirement to destroy Soviet confidence in a first strike against our silos. It is a relatively simple system which fires a large number of small conventional projectiles. This swarm of projectiles forms a barrier against a warhead approaching a U.S. missile silo, engaging the warhead about a mile above the target. At the risk of oversimplifying, this might be described as a computer-aimed shotgun defense at short range. It could also be described as "dynamic hardening" instead of as an antimissile system. If deployed to intercept only the first hostile warhead approaching a silo, it would cost $2–3 million per defended silo. If it is to intercept a second warhead, the costs increase to about $5 million per silo. If the fully developed GAU-8 rapid-fire cannon system were adapted for point defense, costs would probably be lower.

The requirement for an initial spaceborne ballistic missile defense system can be met by using off-the-shelf hardware to create a multiple-vehicle orbiting system. This system would deploy nonnuclear kill vehicles to destroy hostile ballistic missiles in the early phase of their trajectories. Enough of these defensive weapon-carrying satellites would be orbited to ensure continuous coverage of Soviet ballistic missile trajectories, including those of SS-20 Eurostrategic missiles and submarine-launched missiles. This system could provide protection to our allies as well as to the U.S.

The multiple satellite deployment permits each satellite to defend itself and several others. It also has the potential for forming the basis of a highly effective and secure command, control, and communications (C^3) system. Since the system makes maximum use of off-the-shelf hardware, it may be the cheapest and quickest available option. This system deployment could begin in perhaps as little as three years and could be fully deployed in five or six years at a minimum cost of some $15 billion.

The most promising possibility for a second-generation spaceborne defense is product improvement of the first-generation system. With the addition of advanced infrared sensing devices, the spaceborne elements can be made capable of attacking individual warheads, not only during the early phase of ballistic trajectory but throughout the trajectory up to reentry into the atmosphere, which in effect provides two layers of spaceborne defenses. This system could be ready for deployment in 1990 at a cost of about a $5 billion add-on to costs of the first-generation hardware.

The requirement for higher-technology space defense systems might also be met by a high-powered laser system on the ground with redirecting mirrors on satellites, or by beam weapon systems deployed in space or in groundbased pop-up installations. These systems were not conceived during the High Frontier studies; rather, they were proposed as elements of High Frontier because they are already being researched. Costs to continue research should probably be increased by about $100 million per year.

Civil defense is a multifaceted endeavor, with deterrence value and cost-effectiveness which sharply increase when used in conjunction with active defenses. We conclude that increased funding for civil defense is required, but only temporarily. In the long run, the active defenses of High Frontier would reduce the need for civil defense funding. The impact of these conclusions on priorities and costs of current civil defense programs was not analyzed.

There is an urgent need to develop a multipurpose, military, manned space vehicle to perform a wide variety of space missions, e.g., inspection of friendly or suspect space objects; protection of satellites and space stations; and adjustment or retrieval of satellites.

One such vehicle is already under active consideration in the Department of Defense. This vehicle is the high-performance spaceplane, or one-man "space cruiser", which largely utilizes available hardware and technology, and which could be operating in several years for less than $500 million.

The immediate answer to improved space transportation is an upgrade of the current Shuttle program, to improve turnaround time and to create an unmanned cargo-only version. At the same time, development work should begin on a vehicle of much higher load capacity. These programs would cost an estimated $6 billion over a 10-year period.

The currently proposed military Space Operation Center should be given high priority and expanded in concept to include provision for "fly-along" industrial/commercial space installations. The space station should be equipped to receive solar-generated electrical power from a prototype-orbiting SPS. A 10-year program to deploy this space station should cost about $12 billion.

Power requirements can be met by a Solar Power Satellite (SPS) in geosynchronous orbit. The SPS system will include a groundbased microwave receiving antenna and conversion subsystem providing 500 megawatts of continuous electrical power—a working prototype of a nonpolluting energy source which may supplant both hydroelectric and nuclear power plants. This pilot system, modified to provide laser-transmitted power to a space station, would cost about $13 billion. The peaceable use of SPS energy, when fully developed, can be of incalculable value to humankind.

The costs of R&D for industrial space applications would probably be borne almost entirely by interested private enterprise, with no more than $50 million per year in government support.

PROGRAM COSTS

The total costs of the High Frontier system over the next five or six years, in constant dollars, might be roughly $24 billion. Through 1990, the total costs in constant dollars would probably be about $40 billion — a figure that compares favorably with what would have been the total cost of MX–MPS in its original configuration. It also compares favorably with the Apollo Moon-landing program, and strikingly so if the inflation rate of the past 12 years is considered.

The real costs of the High Frontier system are even lower, when we consider possible tradeoffs in programs obsoleted or lowered in priority by the existence of an effective strategic defense. For example, the billions now earmarked for superhardening of existing missile silos, and for deploying more complex point defenses, need not be spent. There are other possible tradeoffs such as repositioning of SAC airfields, reducing the urgency of upgraded nuclear forces in Europe; command, control and communications improvements, and so on.

Finally, there is a reasonable chance for sizeable cost offsets from industry and allied participation in the most expensive aspects of the High Frontier — the nonmilitary aspects. This is especially true if we make a vigorous effort to tap solar energy with SPS. Several nations have already stated their willingness to help in such an SPS effort and such nongovernment support would further reduce the real costs.

In any case, costs to the U.S. taxpayer will certainly be lower in implementing the High Frontier system than in supporting other security programs such as the various expensive schemes for deployment of the MX missile. The High Frontier approach, therefore, cannot be characterized as unrealistically expensive.

IMPACTS

The mere announcement of a bold new U.S. initiative along the lines of the High Frontier system would have beneficial impacts at home and abroad. The *fulfillment* of the urgent requirements, in both the military and nonmilitary dimensions previously itemized, would have even more far-reaching impacts. These impacts might best be characterized as military; political; economic; and foreign.

The impact of the High Frontier system, on the purely military-strategic side, would be to move us away from the unstable arena of MAD to one of Assured Survival—a much more stable condition. We would provide answers to U.S. and allied security problems; answers that would *not* involve the amassing of ever-larger stockpiles, and ever-more expensive deployments, of nuclear weapons.

By creating a proper balance between strategic offense and strategic defense, we broaden our options for strategic retaliatory action. A great deal of the counterforce, damage-limiting function of our strategic forces can be shouldered by our defensive systems. Cruise missiles become a more attractive option in a new setting that includes defenses against ballistic missile attack.

Perhaps more important to our military efforts as a whole, the High Frontier system would restore the traditional U.S. military ethic. The military man's role as defender of our country has always been the tie that has bound him to our supporting citizenry. Strategies of the recent past, such as MAD, deny that role and have seriously weakened that bond. We could greatly reduce problems in all facets of U.S. security through commitment to this new strategy which is consistent with the military rationale of the average U.S. citizen.

The political impact of High Frontier could be a very healthy one, since the potential public support of this system is enormous. If the military and nonmilitary aspects of the system are harnessed together, broad segments of the U.S. public are likely to rally in support. Recent elections have demonstrated a widespread desire for improved defenses. We have a remarkably large support base, primarily among younger people, in the form of space enthusiasts. And there is general public disillusionment with the doctrines and strategies of the past.

The High Frontier effort would even convert or silence some of the conventional opponents of defense efforts and of technological innovations. It is much harder to oppose *nonnuclear, defensive* systems than *nuclear offensive* systems—and with good reason. It is impossible to argue effectively for a perpetual balance of terror, if it can be negated by a new policy. It is hard to make environmentalist cases against space systems, particularly with respect to satellite-generated energy for Earthbound use. The SPS power plants, for the first time in history, can provide abundant energy to us all without the polluting side effects which have always plagued power generation.

Even those naysayers whose basic concern is disarmament will be hard-pressed to make a case against High Frontier—the ABM treaty notwithstanding. It is not necessary to abrogate the ABM treaty to commit to High Frontier programs. The High Frontier spaceborne defenses fall into the category described in the treaty as "systems based on other principles" which are "subject to discussion" with the Soviets. Point defense systems can also be selected which are so different from ABM systems as defined in the treaty that they too could be considered as outside the treaty. Indeed, some silo defense systems can be considered as "dynamic hardening"—a substitute for reinforced concrete—rather than an ABM. Further, the U.S. could propose any amendments deemed necessary to the ABM treaty to accommodate strategic defensive decisions.

A U.S. commitment to the High Frontier system does not necessitate rejection of arms negotiations with the Soviets. It does, however, mean that future negotiations would proceed on a different philosophical basis. Rather than continue to pursue agreements which attempt to perpetuate the MAD balance of terror, our negotiating efforts would be dedicated to achieving a stable world of Mutual Assured Survival.

Presuming that we embark on the High Frontier, we should not object or be surprised to see the Soviets follow suit—if indeed they are not already developing such a system in secret. While military strategists may grumble to see an adversary succeed in any new arena, the movement of both adversaries toward defensive strategies is ultimately a movement toward stability.

There can be little doubt that a strong commitment to the High Frontier by the U.S. would have highly beneficial economic impacts. Some of these impacts will affect the U.S. economy in the near term, primarily stimulated by investment in high-technology sectors of industry—and probably by a general upswing in confidence. We estimate an increase of 200,000 jobs in the near term as a result of a strong commitment to space. Longer-term impacts will depend on the rate at which industrial applications are realized, and on as-yet-unpredictable spinoffs from the space effort.

One example of commercial space application is already paying its way very well. Space communications is a $500-million-per-year enterprise and is growing rapidly. By 1990 it should become a multibillion-dollar-per-year industry. Another spinoff from space technology is the miniaturized computer which, in all its guises from timepieces to office word processors, has already become a multibillion-dollar-per-year industry in the very teeth of an economic recession. Without speculating on which spinoff is likely to become our next economic strength, we can cite the fact that space technology programs have provided some of our most striking commercial successes of the past generation. As other industrial applications of space technology are realized, the total revenues from space industries may reach levels of tens of billions of dollars-per-year by the year 2000.

Some of the most beneficial economic impacts of a strong High Frontier program are indirect and unquantifiable. The demand for highly skilled workers is certain to have a positive impact on our education system and on the labor market. New products, tools, and services will be required by an expanding space effort. We will intensify research efforts.

Overall, the economic benefits of a vigorous U.S. commitment to space exploitation—for both security and industry—are potentially very great. Yet in the details, these benefits are no more predictable today than were the future economic benefits of aviation in the 1920s.

The positive political impacts of High Frontier in the U.S. will probably be reflected in positive foreign impacts among our allies. Our announcement of a commitment to the High Frontier system could have a strong pacifying effect on the current highly disruptive "antinuclear" or "peace" movements in Europe. A bold U.S. strategic defensive initiative would certainly bolster the morale of pro-U.S. groups. The High Frontier can, in fact, become a new cement for Free World alliances, making them global rather than regional.

A shared U.S.-Allied commitment to the harnessing of solar power from space could have highly beneficial impacts on foreign relations. If the prospects were good for abundant future energy supplies, independent of fossil fuels and their locations, the industrialized West would no longer be overdependent on the oil and gas producing countries. Further, the prospects for overcoming the refractory problems of underdeveloped nations could have a beneficial impact on the attitudes of the Third World.

As for the Soviets, their initial reaction is easily predictable as hostile. They have already moved to counter the U.S. potential to adopt available military space options. They have introduced in the United Nations (and garnered some support for it among our allies) a new treaty which would ban *all*—not just nuclear— weapons in space. Meanwhile, evidence mounts that the Soviets are already in violation of their own cynical proposition. We can expect an extraordinarily strong Soviet propaganda effort against a U.S. commitment to the High Frontier, including threats of counteraction. However, in both particulars Moscow will find, for substantive reasons, an attack on the High Frontier concepts much more difficult to conduct or to rationalize than past anti-U.S. campaigns.

MANAGEMENT IMPERATIVES

We conclude that management of the High Frontier effort must be organized around the imperatives of timeliness and streamlining of bureaucratic processes. Time is critical in any commitment to the High Frontier, especially with regard to the military systems. If we cannot change the adverse trends in the military balance quickly, we may not be able to change them at all. If we do not move quickly to secure space for promising industrial development, we may later be denied the opportunity.

There are no fundamental technical obstacles to meeting the military and nonmilitary objectives of High Frontier. We can close the window of vulnerability in two or three years, and negate the menace of MAD in five or six years. We can lower the costs of men and materials in space; establish a permanent manned presence in space; and open the door to enormous economic advantages, all in ten years. However, we can do these things only by initially selecting systems that utilize off-the-shelf technology to the maximum, and by instituting special management and procedural arrangements for rapid deployment of High Frontier elements. Using known, tested technology, we can avoid many of the long delays imposed by research and development. By special management arrangements, we can avoid bureaucratic hurdles which have been inserted into our weapons acquisition processes over the past 15 years. Time is money, and literally billions can be saved by cutting acquisition times.

In 1956, President Eisenhower gave the go-ahead on a concept for a ballistic missile-firing submarine. That concept involved far more technological unknowns than do the High Frontier options. In 1960, just 47 months later, the first Polaris put to sea. In 1962, President Kennedy announced the objective of landing a man on the Moon. Seven years later, this astonishing feat was accomplished.

Today, even a new fighter aircraft takes 13 years or more from concept to acquisition; and decades of delay are predicted for space developments. These protracted processes cause costs to soar astronomically. This sad state of affairs exists not because Americans have become technologically inept, but because we have, over the years, constructed a complex and multi-layered bureaucratic system in the Executive Branch and in the Congress—and this system simply cannot produce quick results. In order to take advantage of our opportunities on the High Frontier, we must—at least for a few years—find a way to short-circuit the bureaucratic institutions and procedures.

The first step is to select—and select quickly—those elements which will best meet the urgent requirements of the High Frontier system. This should be done by a Presidential Systems Selection Task Force composed of prominent and properly qualified individuals.

To provide overall guidance to the High Frontier effort, a National Space Council should be appointed with representation from the involved departments and agencies of the Executive Branch, Congress, and industry. Its function would be to insure full coopera- tion and fast action by all involved branches of govern- ment and private industry. Its chairman should be the Vice President.

The actual coordinating and expediting of the selected High Frontier programs should be the responsibility of a chief operating officer heading up a Consolidated Program Office. This officer should be assisted by special project officers within the departments and agencies charged with acquiring the first-generation High Frontier system elements. The management team should insure *individual* rather than *committee* respon- sibility for decisions; a minimum of Executive and Congressional staff review; and specified or "fenced" funding for High Frontier programs.

This management system should, unequivocably, be temporary. It should go out of existence upon achieve- ment of its objective, i.e., acquisition of the first- generation system. As results are obtained, all respon- sibility for the operations, maintenance, and further growth of space systems should return to the cogni- zance of the appropriate agencies — Defense and NASA. There is no need to create a new permanent layer of bureaucracy.

These are the essentials of the High Frontier system. They are discussed in much greater detail in later chapters of this book. We believe that the change in strategy represented by the High Frontier system sup- ports a U.S. policy statement which follows.

PROPOSED STATEMENT OF U.S. POLICY

The United States and its allies now have the combined technological, economic, and moral means to overcome many of the ills that beset our civilization. We need not pass on to our children the horrendous legacy of "Mutual Assured Destruction", a perpetual balance of terror that can but favor those most inclined to use terror to bring down our free societies. We need not succumb to ever-gloomier predictions of diminishing energy, raw materials, and food supplies. We need not resign ourselves to a constant retreat of free economic and political systems in the face of totalitarian aggressions. The peoples of the Free World can once again take charge of their destinies, if they but muster the will to do so.

In April of 1981, the Space Shuttle Columbia made its dramatic maiden voyage into space and back safely to Earth. This event was not merely another admirable feat of American space technology. It marked the advent of a new era of human activity on the High Frontier of space. The Space Shuttle is a development even more momentous for the future of mankind than was the completion of the transcontinental railway, the Suez and Panama Canals, or the first flight of the Wright Brothers. It can be viewed as a "railroad into space" over which will move the men and materials necessary to open broad new fields of human endeavoɪ in space, and to free us from the brooding menace of nuclear attack.

This is an historic opportunity—history is driving us to seize it.

A few thousand years ago, man's activities—his

work, his commerce, his communications, all of his activities including armed conflict — were confined to the land.

Eventually man's technology and daring thrust his activities off the land areas of the continents and into the coastal seas. His work, commerce, communications, and military capabilities moved strongly into this new arena of human activity. Those nations that had either the wit or the luck to establish the strongest military and commercial capabilities in the new arena reaped enormous strategic advantages. For example, the Vikings, although never a very numerous people, became such masters of the coastal seas that their power spread from their homes in Scandinavia over all the coasts of Europe and into the Mediterranean Sea, up to the very gates of Byzantium.

At the beginning of the 16th century, after the epic voyages of men like Magellan and Columbus, human activity surged onto the high seas. Once again, the nations that mastered this new arena of human activity reaped enormous strategic rewards. First Spain and Portugal utilized their sea power to found colonies and to solidify their strength in Europe. Later, Great Britain, with an unsurpassed fleet of merchantmen and fighting ships, established a century of relative peace which we remember as Pax Britannica.

In the lifetime of many of us, man's activity moved strongly into yet another arena, the coastal seas of space — the air. And once again the nations which quickly and effectively made use of this new arena for commerce and defense gained great advantages. As Americans we can take pride that the greatest commercial and military successes in aviation have been achieved by our nation.

But today, following the epic voyages of our astronauts to the Moon and our unmanned space probes exploring the rings of Saturn and beyond, we find man's activities moving strongly into yet another new arena—the high seas of space. Already the United States and other major nations, including the Soviet Union, are making huge investments in space. Much of our communications, intelligence, weather forecasting, and navigation capabilities are now heavily dependent on space satellites. And, as history teaches us well, those nations or groups of nations that become preeminent in space will gain the decisive advantage of this strategic "high ground".

We must be determined that these advantages shall accrue to the peoples of the Free World; not to any totalitarian power. We can improve the Shuttle, our railway into space, placing space stations at its terminals and sharply reducing the cost-per-pound of material put into space. We can thus open the doors of opportunity to develop entire new space-based industries, promising new products and new jobs for our people on Earth. We can eventually create the means to bring back to Earth the minerals and the inexhaustible solar energy available in space. By doing so, we can confound the gloomy predictions of diminishing energy and material resources here on Earth. This will not only enhance the prosperity of the advanced, industrialized nations of our Free World, but will also provide the means to solve many of the hitherto intractable problems of the developing countries.

Further, we can place into space the means to defend these peaceful endeavors from interference or attack by any hostile power. We can deploy in space a purely defensive system of satellites using nonnuclear *weapons which will deny any hostile power a rational option for attacking our current and future space vehicles or for delivering a militarily effective first strike with its strategic ballistic missiles on our country or on the territory of our allies. Such a global ballistic missile defense system is well within our present technological capabilities and can be deployed in space in this decade, at less cost than other options that might be available to us to redress the strategic balance.*

We need not abrogate current treaties to pursue these defensive options. A United Nations treaty prohibits the emplacement of weapons of mass destruction in space, but does not prohibit defensive space weapons. The ABM Treaty requires discussion among Soviet and U.S. representatives of any decision to proceed with defensive systems "based on other principles" such as space systems. We should initiate such discussions and propose revisions, if necessary, to the ABM Treaty.

Essentially, this is a decision to provide an effective defense against nuclear attack for our country and our allies. It represents a long-overdue concrete rejection by this country of the "Mutual Assured Destruction" theory which held that the only effective deterrent to nuclear war was a permanent threat by the United States and the Soviet Union to heap nuclear devastation on the cities and populations of the other. The inescapable corollary of this theory of MAD was that civilian populations should not be defended, as they were to be considered hostages in this monstrous balance-of-terror doctrine. The MAD doctrine, which holds that attempting to defend ourselves would be "destabilizing" and "provocative", has resulted not only in the neglect of our active military and strategic defenses and our civil defense; it has also resulted in the near-total dismantlement of such strategic defenses as we once had.

For years, many of our top military men have decried the devastating effect the MAD theory has had on the nation's security. In fact, our military leaders have, over the years, denied its validity and tried within the limits of their prerogatives to offset its ill effects. But those effects are readily evident. The only response permitted under MAD to increased nuclear threats to the United States or to its allies was to match those threats with increased nuclear threats against the Soviet Union. Further, a U.S. strategy which relied at its core on the capability to annihilate civilians and denied the soldier his traditional role of defending his fellow citizens has had a deleterious effect on the traditional American military ethic, and on the relationship between the soldier and the normally highly supportive public.

This legacy of MAD lies at the heart of many current problems of U.S. and allied security. We should abandon this immoral and militarily bankrupt theory of MAD and move from "Mutual Assured Destruction" to "Assured Survival". Should the Soviet Union wish to join in this endeavor—to make Assured Survival a mutual endeavor—we would, of course, not object. We have an abiding and vital interest in assuring the survival of our nation and our allies. We have no interest in the nuclear devastation of the Soviet Union.

If both East and West can free themselves from the threat of disarming nuclear first strikes, both sides will have little compulsion to amass ever-larger arsenals of nuclear weapons. This would most certainly produce a more peaceful and stable world than the *one we now inhabit. And it would allow us to avoid leaving to future generations the horrendous legacy of a perpetual balance of terror.*

What we propose is not a panacea which solves all of the problems of our national security. Spaceborne defense does not mean that our nuclear retaliatory capabilities can be abandoned or neglected. The United States would still maintain strategic offensive forces capable of retaliation in case of attack. The Soviets, while losing their advantage in first-strike capabilities, would still be able to retaliate in case of attack. Nor does our approach to the strategic nuclear balance eliminate the need to build and maintain strong conventional capabilities.

We Americans have always been successful on the frontiers; we will be successful on the new High Frontier of space. We need only be as bold and resourceful as our forefathers.

CHAPTER II
STRATEGY

The High Frontier study group intensively reviewed the spectrum of threats facing the United States and its allies, the global problems associated with those threats, and the options available to meet them. We urgently recommend the adoption of a new national strategy of Assured Survival to replace MAD. Assured Survival can be achieved by using U.S. technological advantages, especially in space, to provide our citizens with long-neglected protection from nuclear attack and to secure space for our long-term economic benefit.

This new strategy would:

- Provide for the defense of the U.S. and its allies against nuclear ballistic missile attack.

- Secure the availability of the vast resources of space to the U.S. and the Free World, while providing for its defense against hostile attempts to deny the use of this great medium for peaceful purposes.

- Capture the imagination and support of the broadest spectrum of Free World peoples, and restore to them a sense of optimism and purpose while peacefully neutralizing the Soviet strategic nuclear menace.

- Require vigorous development of the economic opportunities available to us on the High Frontier of space—opportunities open to us because of hard-won advantages in space technology. These economic developments would materially benefit the prosperity of industrial nations of the Free World and would also address the presently intractable problems of the lesser developed nations.

We believe that such a strategy is sound, technologically feasible, well within our capabilities, fiscally within our means, and likely to engender strong public support at home and abroad. The High Frontier strategy could extract us from our current, strategically flawed, positions in several arenas. A brief recap of our problems is in order.

For more than a decade, we have cultivated unsound policies and doctrines based on illusory notions concerning the nature of threats to U.S. security and vital economic interests. We are now reaping a bitter harvest from that cultivation. There is a serious and growing Soviet advantage in strategic power which the U.S. cannot counter except by a threat of retaliation which involves national suicide. The preponderance of Soviet conventional power vis-a-vis the U.S. and its allies is also severe and growing. It can no longer be counterbalanced, as it has in the past, by a credible threat to bring higher-technology U.S. weaponry to bear. The Soviet Union is increasingly successful in its use of propaganda and the application of direct or indirect military power to disrupt our alliances, and to force Marxism on underdeveloped countries. This Soviet success now threatens the continuing availability of raw materials critical to the industrialized West. To take a specific example, the West is dangerously dependent on diminishing crude oil supplies located in areas threatened by Soviet military or manipulative political power. The Soviet adventure in Afghanistan is an obvious thrust in the direction of crude oil supplies.

Moreover, the U.S. alliance system is in serious disarray. It suffers from a lost sense of purpose and a perception of decline in U.S. power and leadership. The Soviet propaganda offensive against U.S. nuclear weapons was designed to persuade Europeans to become neutral, and is increasingly effective.

Fortunately, this litany of Western woes is not the whole picture. The USSR has generated problems of its own. It is suffering the strains of imperium in Poland and Afghanistan. Inside the USSR, the Kremlin is faced with a small but growing group of dissidents among its elites, and with the prospects of Great Russians becoming an ethnic minority (see, for example, D'Encausse's *Decline Of An Empire*, Newsweek Books, 1979). Further, the Soviet Union's huge commitment of resources to its military machine over the past 15 years has impaired its already chronically deficient general economy.

This combination of threat and opportunity provides the U.S. with an historic but transitory opportunity to change the world for the better.

OUR STRATEGIC OPTIONS

There are two basic options available to meet the military strategic challenge: an incremental modification of our existing strategy, or a bold new initiative.

We could attempt to improve the situation by merely modifying the basic strategies of the past and by adding resources to the programs designed to support that strategy. This would entail incremental changes in our strategy, and in our military programs along rather predictable lines. For example, MAD would remain the unspoken (although frequently denied) cornerstone of our force structure; *but* we would modify our offensive forces to insure a somewhat higher level of destruction to the USSR in the event of attack, while continuing to ignore active strategic defense or effective civil defense. We would continue to rely on arms-control treaties, past and future, as the answer to national security at reasonable cost; *but* we would try to get tougher with the Soviets at the bargaining table. Finally, we would remain content with the old concepts of "parity", "essential equivalence", etc.; *but* we would decry numerical imbalances and would add billions of dollars to military programs presumed to close gaps between the U.S. and the Soviets in current weapons and forces.

This incremental approach will almost certainly fail us — strategically, politically, and economically — for several reasons. For one thing, attempts to close the arms gap with the Soviets by adding more hardware of current technology (missiles, aircraft, ships, tanks, etc.) play to their long suit. The Soviets are already producing these items at a very high rate, far surpassing current production rates of the U.S. and its allies. With a universal draft, they are also able to man that hardware effectively. In a contest for sheer military mass, the Soviets will probably be even farther ahead of us in four years than they are today (see Figure 1). Our industrial mobilization base is grossly inadequate for such competition in arms production, and in reality is incapable of closing the hardware gap within an acceptable timeframe.

Secondly, we would court political failure. Mere modifications to previous strategy (MAD, disarmament, detente), without the necessary changes in real capabilities, would confuse rather than clarify public perceptions of U.S. policy at home and abroad, as was evidenced in the public's reaction to Carter's Presidential Directive 59 (issued in 1980). Eventually, the lack of results and the high costs of the incremental approach would force arms-limitation talks back into a predominant position in U.S. security policy. The Reagan Administration would begin to sound more and more like its predecessor, and the mandate for strong defense and foreign policy would dissipate.

This incremental approach would also be an economic failure, because the inclination of the public and hence of Congress to support very high defense expenditures is a diminishing asset. The incremental approach does little to sustain public support for a strong national defense. Further, it creates the grave danger of a severe backlash against current proponents of increased defense expenditures if, four years hence, there is no perceptible, favorable change in the U.S.-Soviet military balance. Our other basic option envisions a new national strategy on the High Frontier. It rejects the MAD doctrine outright, faces squarely the failure of arms-control efforts to date, and end-runs current Soviet strategic advances with superior technology.

In this bold approach, we would replace MAD with Assured Survival, through emphasis on strategic defense which maximizes the use of known space tech-

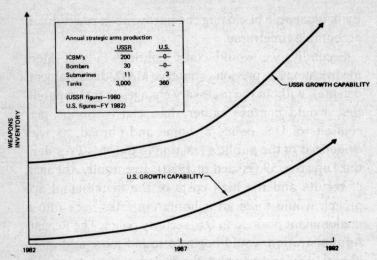

Figure 1. Incremental Approach

nology and available point-defense options. This would create, in the shortest possible time, an effective Free World defense against Soviet nuclear attack or blackmail. We would adopt those systems most readily adaptable to foreseeable technological advances, such as laser and other beam weaponry. We and our allies would address the broadest spectrum of global problems by pursuing parallel nonmilitary programs to open space for promising industrial and commercial development. We would initiate near-term nonspace programs compatible with the new strategy, which would enhance the value of mid- and longer-term space programs. We could hold total costs at or below current budget programs. We could achieve a measure of this new deterrence long before the entire system is deployed because, with deployment of the first GBMD elements, we would greatly reduce the Soviet assurance of a first strike's effectiveness. Finally, we could create a management structure designed to overcome the long lead times inherent in new systems development and acquisition procedures.

The bold new High Frontier approach offers strong possibilities for avoiding the pitfalls of the incremental option, while providing answers to the broad range of strategic problems that beset the U.S. and its allies. First, we can avoid leaving to future generations the sorry legacy of a perpetual MAD balance of terror, which in the end must favor the side most inclined to *use* terror. Also, the High Frontier moves our contest with the USSR from an arena where the Soviets have the advantages to one in which the U.S. has the advantages. Further, ours is a truly international, rather than a merely *military,* strategy. It provides—in addition to security—promising solutions to economic problems (including the energy crunch), to the problems of development in the Third World, and to the erosion of spirit in the West. Our new strategy can also reverse the alienation of the public toward its military establishment by making the military effort understandable to the average U.S. citizen and compatible with his own interests. At the same time, the defensive nature of the new strategy will have the effect of disarming much of the antidefense/antitechnology sentiment in the Free World by offering a viable alternative to the continuous growth of destructive nuclear-weapon inventories. Finally, the High Frontier strategy can accomplish all these things in less time, with less money, and with far more popular support than any available alternatives.

In our efforts to examine all dimensions of this new strategy, we have considered international ramifications, including Soviet reactions and effects on existing treaties, cost and macroeconomic effects, management options to implement the programs, and the near-term, nonspace collateral support needed for the new strategy. All of these factors are detailed in succeeding chapters of this book.

The results of the High Frontier studies, of course, cannot be as thorough as would be a similar effort by the government; but the results appear sufficiently conclusive to support a change in U.S. strategy.

CHAPTER III
THE MILITARY DIMENSION

The dimensions and severity of the military imbalance between the U.S. and the Soviet Union, as previously noted, need no further elaboration. Suffice it to say that the Soviets can ensure many decades of global dominion if they are successful in their current effort to add military domination of near-Earth space to their already massive strategic power. This prospect alone demands a vigorous effort on the part of the U.S. to maximize our current technological advantages to open this new frontier in defense of the Free World.

The imperative to meet this Soviet challenge in space is reinforced by the paucity of options available to us to meet the current threat by other means. In the recent debate over deployment modes for the MX missile, it became clear that a search for invulnerable offensive weapons is fruitless. While some modes are less vulnerable (e.g., submarines), none can be reliably predicted to remain invulnerable for more than a few years—especially when potential military systems *in space* are postulated as part of the Soviet arsenal.

Further, we find that an attempt to redress the imbalances by incremental add-ons to land, sea, and air components of U.S. forces requires very large and politically difficult expenditures. While perfectly reasonable arguments can be sustained for such expenditures (e.g., historical percentages of GNP devoted to defense), and perhaps should be persuasive, the fact is that a tailchase of the Soviets in producing current technology weapons would probably be a strategic failure. The U.S. would start with a very low arms production rate from a seriously diminished arms production base, while the Soviets would proceed from a formidable arms production rate and base.

Any search for a "technological end-run" on Soviet military advantages leads inexorably to space. While there are promising technological innovations possible in current land, sea, and air weaponry, these innovations are essentially product improvements unlikely to cause more than small vernier changes in the overall strategic balance. It is in the area of space technology where the U.S. advantage can be decisive. While the Soviets lead us today in the *application* of space technology to military capabilities, we have a strong lead in *potential*. The Space Shuttle is the most obvious of our advantages, but our fundamental advantage lies in our ability to miniaturize and therefore achieve superior capabilities per pound of materials put into space.

Careful examination of U.S. military options in space leads to the firm conclusion that we must make a priority effort to reestablish strategic defense. The only way that we can effectively shift the strategic framework of U.S. and allied thinking away from the MAD doctrine, to Assured Survival, is to deploy a global ballistic missile defense (GBMD). This can be effectively accomplished only in space. Partial withdrawal from the MAD doctrine is possible through deployment of point ABM defenses and renewed attention to civil defense. However, a true break with the all-offense, no-defense approach can be *decisive* only when it includes an effective spaceborne defense element. This is true because space offers the only potential for general, global strategic defense of the entire Free World at reasonable cost.

We are well aware that there have been cogent denials among military spokesmen that MAD has, in fact, been official U.S. doctrine; some have attempted within the limits of their prerogatives to avoid the "mutuality" of Assured Destruction. The salient fact, however, is this: since the mid-sixties, the MAD doctrine has been sufficiently pervasive to prevent any serious attention to strategic defense options. All responses to increased strategic nuclear threats have been in terms of increased U.S. strategic nuclear *offense*.

We cannot, of course, allow our strategic offense to decay while we concentrate on strategic defenses. Political and military realities demand that we adopt balanced, strong, and mutually supportive offensive and defensive forces.

The Case for Strategic Defense

Certain articles of faith of the no-strategic-defense school of thought must be addressed and discarded. In the past, considerations of strategic defense options have been attacked and defeated with several false conclusions. Opponents argued that strategic defense systems are useless in the nuclear era unless they are impermeable, i.e. *perfect*; that they are of little value if they can also be attacked along with the targets they defend; that they are impossibly expensive;

and that they are destabilizing because they cause the opponent to believe we contemplate attack.

The notion that the strategic defense of populations and homelands is somehow "bad" and "destabilizing," and that mutually retaliatory strategic offense forces are "good" and "stabilizing," is the peculiar "wisdom" of the post-1945 period. Indeed, this particular complex of assumptions has been discernible in British and American strategic debate since the early 1900s, especially in the period just before World War II. At that time, a high-level group—the Air Defense Research Committee—was established by the British government to look at the prospects for defending Britain from the German Luftwaffe. The question then was whether the Royal Air Force would allocate its funds to a "retaliatory" bomber force, in line with one school of analysis, or whether it would build up a homeland defense force of Spitfire and Hawker Hurricane interceptors, as well as radars and civil defenses. The threat of annihilation that the British government perceived at that time was parallel to what we see today in Soviet nuclear ballistic missiles. The British anticipated heavy Luftwaffe use not only of incendiaries but also of gas bombs against British cities. Some saw this threat as impossible to defend against with available or foreseeable technology. Fortunately for the Western democracies, the British government came down on the side of the interceptor force, the civil defense air raid wardens, and the radars that together won the Battle of Britain. Unlike the American government 30 years later, it did not embark on a wholesale ideological policy excursion in the direction of *mandatory* vulnerability of its own homeland.

The rejection of the lessons of the Battle of Britain,

and the demise of serious U.S. consideration of strategic defense, were signaled in a speech by then-Secretary of Defense Robert McNamara on September 19, 1967. He specifically called for an impermeable defense or no defense at all:

> . . . it is important to understand that none of the (ABM) systems at the present or foreseeable future state of the art would provide an impermeable shield over the United States. *Were such a shield possible, we would certainly want it — and we would certainly build it. . . . If we could build and deploy a genuinely impenetrable shield over the United States, we would be willing to spend not $40 billion but any reasonable multiple of that amount that was necessary.* The money in itself is not the problem: the penetrability of the proposed shield is the problem.

(Emphasis added. Source: *Department of State Bulletin*, October 9, 1967.)

The original validity of the antidefense arguments can be readily challenged. Their validity is even more suspect in light of strategic defensive options available today, nearly a generation later.

With regard to invulnerable defenses, there never has been nor ever will be a defensive system that could meet such criteria. Such perfectionist demands ignore the purposes of defenses and the effects of strategic defense on deterrence. Defenses throughout military history have been designed to make attack more difficult and more costly — not *impossible*. Defenses have often prevented attack by making its outcome uncertain. General Grant put a cavalry screen in front of his forces not because the cavalry was invulnerable to Confederate bullets, nor because he thought the cavalry could defeat General Lee, but because he did not want the battle to commence with an assault on his main forces or his headquarters.

It is this same military common sense that must prevail in our approach to strategic defenses today. Given the drastic consequences of a failed nuclear attack on an opponent, the critical military task is to keep a potential aggressor *uncertain of success*. If he can be made certain of failure, so much the better. In the absence of defenses, the Soviet military planner has a rather straightforward arithmetic problem to solve to be quite sure of the results of a disarming strike against all locatable U.S. strategic weaponry— ICBM sites, airfields, and submarine bases. His problem is simply to ensure that he can deliver two warheads of current size and accuracy against each such target. If, on the other hand, the Soviet planner must consider the effects of a strategic defense (especially a spaceborne defense which destroys a portion of the attacking missiles in the early stages of their trajectories), he is faced with a problem full of uncertainties. He does not know how many warheads will arrive in the target areas nor—even more crucial—*which ones* will arrive over which targets. This changes the simple arithmetic problem into a complex calculus full of uncertainties; *and such uncertainties are the essence of deterrence.*

Strategic defenses are eminently practicable and by no means impossibly expensive, if the programs involved are not required to meet unrealistic standards of perfection or incredible postulated threats. A cursory review of combinations of spaceborne defenses, landbased ABMs, and civil defense—while by no means *definitive* as to costs—indicates that a layered strategic defense system (see Figure 2) can be devised which is of decisive strategic importance and relatively inexpensive when compared with some previous offensive systems.

One attractive option open to us is to create a space-borne ballistic missile defense quickly, using essentially off-the-shelf technology. We provide as an example of such a system the Global Ballistic Missile Defense (GBMD) system described in detail in Chapter VI.

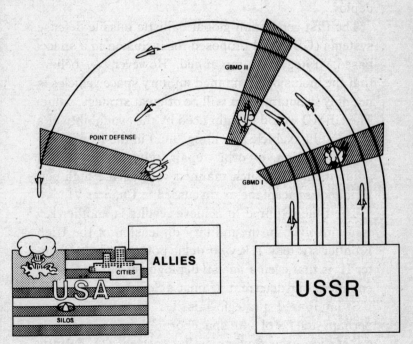

Soviet strategic nuclear missiles are attacked in early, mid, and terminal phases of their trajectories.

Figure 2. Depiction of Layered Defense

This system is a multiple, unmanned satellite system which employs nonnuclear kill vehicles (already largely developed) to strike Soviet ballistic missiles in the early stages of their trajectories. It is a relatively rugged and uncomplicated system which can readily accommodate improvements or enhancements which might prove attractive in the future.

Clearly, the deployment of a defensive system in space would sharply reduce the vulnerability of operating U.S. space systems (communications satellites, intelligence gatherers, and Shuttle), as well as future military and civilian systems which we may decide to deploy.

The first-generation global ballistic missile defense systems (GBMD-1) proposed for inclusion in a space-based strategy are unmanned. However, we believe that the inclusion of manned military space vehicles is not only inevitable, but will be of great strategic value. The GBMD would be enhanced by the availability of a space utility vehicle providing "man-in-the-loop" security, inspection, on-orbit repair, refurbishment, and adjustment. One such manned vehicle, the high-performance spaceplane, is discussed in Chapter VI.

The time required to achieve results is another key issue involved in the military dimension of the High Frontier strategy. A key strategic point, made in Chapter II, is that even a *partial* deployment of our GBMD enhances our deterrent against a Soviet first strike on U.S. landbased strategic missiles. In essence: when perhaps ten % of our spaceborne defense system is on station, the Soviets will experience a definite decrease in their ability to calculate the results of a first strike.

The time required to bring a spaceborne GBMD to

bear on the strategic imbalance depends heavily on
our adoption of a management system capable of
accelerating decision and procurement times which
are now creating intolerable time lags in weapon system acquisition (see Chapter IX).

Whatever the selection of spaceborne systems might
be, we believe it is also necessary to provide more
immediate groundbased ballistic missile point defenses
to hedge our bets against delays beyond the critical
period of the "window of vulnerability", a period in
the 1980s when the Soviets could knock out the bulk
of U.S. groundbased ICBMs in a first strike. A groundbased point defense layer would, as we have said previously, further complicate the problems faced by Soviet
strike planners. The criteria for such point defenses
should include low cost and deployment within two or
three years. Its minimum essential function is to prevent a *confident* Soviet first strike against U.S. landbased
strategic missile silos. Such systems are available, and
some examples are discussed in Chapter V.

It is important to note that such simple ABMs have
not been the focus of attention of past U.S. development.
Much more sophisticated and expensive groundbased
systems would be required to provide an effective
strategic defense in the absence of an effective spaceborne GBMD "filter". The low-technology, low-cost
systems can provide near-term uncertainty in the minds
of Soviet planners; uncertainty that is necessary to
reestablish confidence in our deterrent. The military
value of these point defense elements will increase
sharply when they need confront only those warheads
that leak through a spaceborne defense.

The military dimension also includes renewed attention to civil defense, which becomes a far more man-

ageable problem in conjunction with active strategic defense (see Chapter VII).

Since spaceborne defensive systems do not cancel out the need for an adequate balance of offensive systems, we have examined some options in that area. We advocate substantial strengthening of our offensive deterrent strength. The requirements to replace our aging strategic bombers, missiles, and missile-launching submarines are certainly not canceled by a new emphasis on strategic defense. However, some currently popular options, e.g., superhardening of silos, appear less effective in strengthening our deterrent than the High Frontier layered defense.

Nonetheless, the existence of effective strategic defense is bound to have some effect on strategic offensive system programs. These impacts will go beyond the inevitable competition for defense dollars. When the layered strategic defense system of the High Frontier is implemented, the damage-limiting function of our strategic forces will become a shared responsibility between defense and offense. It will no longer be the sole function of counterforce offensive systems. While this does not completely remove the requirement for counterforce capabilities in our TRIAD systems, it *does* affect the rationale for urgency and priority. Obviously, the defensive systems that must be developed (on Earth and in space) must receive higher priority relative to offensive systems than in the past.

SURVIVABILITY OF SPACE SYSTEMS

A key issue which arises with regard to all space weaponry—or for that matter, space hardware in general—is survivability. Space vehicles are nearly invulnerable to some threats such as attacks by troops, terrorists, or saboteurs armed with a wide variety of weapons. On the other hand, all space vehicles or installations now in operation or contemplated for the future are subject to attack by a determined enemy with an adequate technological base. This is true in any military endeavor. In fact, given the special vulnerabilities of space vehicles, if vulnerability to attack were the overriding consideration, there would not now be (nor would there *ever* be) any important space hardware in orbit.

Much of our current understanding of space-vehicle vulnerability is based on the characteristics of our deployed individual, highly complex satellites, incapable of self-defense. The concern about vulnerability is much diminished when one considers a multiple-satellite military system, with each satellite capable of defending itself and many of its companions. Nothing is invulnerable to attack—space vehicles or otherwise—and all else being equal, the more complex the machine, the more vulnerable it is. Complex machinery on Earth has serious vulnerabilities not shared by systems in space. While space systems cannot now be hidden from sight, encased in thick walls of steel and cement, or protected by barbed wire and soldiers, they are not subject to attack with ordinary weapons. It is highly unlikely that they could be disabled or rendered ineffective by natural events. In this regard, current space systems—although vulnerable in orbit—are more vulnerable in their Earthbased links.

While relatively immune to many threats, current space systems have some special, serious, and thus far inadequately countered threats to their survivability.

These threats derive from certain characteristics of current space systems:

1. They are physically fragile.

2. They are highly complex and delicate.

3. They cannot be hidden from the view of ground sensors.

4. They usually travel in fixed or nearly fixed orbits, making their exact locations highly predictable.

5. They travel at very high relative speeds, which makes the impact with even a very small object highly destructive.

6. They operate in a vacuum which does not attenuate or scatter various forms of energy directed at them.

7. They are devoid of means for active self-defense.

Some technical actions have been taken by the U.S. to reduce these vulnerabilities. Electronic components have been protected against *far distant* sources of radiation (e.g., electro-magnetic pulse — EMP — produced by nuclear explosions), but they remain vulnerable to less distant threats. Some vehicles are capable of changing orbits to avoid certain types of hostile action. Still, our space systems remain essentially vulnerable, viable only as a result of Soviet tolerance which derives in large part from political considerations. We must calculate that in time of war, most of our currently undefended satellites would be destroyed in minutes, the rest within hours — a calamitous loss when one considers the extent to which we rely on our satellite systems for defense purposes.

Given the characteristics of current satellite systems, we have noted a number of postulated attack modes and analyzed them individually. Some of the attack modes are possible for the Soviets to employ now. The USSR has a rather primitive but potentially effective antisatellite system for single attack against a satellite in certain orbits. Farther, they can now fire nuclear missiles at each of our satellites which could almost certainly destroy them. Within a few years, and within their *available* technology, the Soviets could probably develop capabilities to shoot down satellites over their own territory with direct ascent, nonnuclear missile attacks. Farther in the future, beam weapons—ground or spacebased—can pose even more serious threats to targets directly overhead.

Postulated attack modes, however, do not constitute real vulnerability problems for a given space system in *wartime* unless they cannot be mitigated or countered by technological or military measures. In *peacetime*, postulated attack modes do not translate into system vulnerability unless they can also pass the test of political credibility.

We examined the survivability problem and its relationship to U.S. options for a GBMD in Chapter I. We also examined current and postulated future Soviet attack modes against U.S. space vehicles, and we found that they fall into two basic categories: peacetime attack and wartime attack. In wartime, the attack could be designed to destroy U.S. systems or impair their utility.

A much larger range of *current* Soviet capabilities could be brought to bear in peacetime. It is technically possible for the Soviets to attack U.S. satellites as they pass over certain areas of the Soviet Union with either direct-ascent missiles or with their antisatellites (ASATs) armed with various nonnuclear destructive payloads. Some of these possibilities are difficult to offset technically without severe weight penalties to U.S. systems. Further, laser weapons currently available to the Soviets could, after numerous attacks on each satellite, wear away protective shields and, in time, destroy them. However, it is extremely doubtful that the Soviets would take such risks in peacetime, since such action would be tantamount to attacking any other element of U.S. defenses. Therefore, otherwise viable space defense options should not be ruled out on the basis of this class of technically feasible but politically incredible threat.

The wartime threat is more critical. The most serious of *current* wartime threats to U.S. space systems are those of nuclear attack, either to destroy or disable, using the effects of radiation from nuclear explosions. Individual U.S. satellites or systems involving small numbers of satellites are more vulnerable to such attacks than are multiple-satellite systems. We already have means to partially shield systems against radiation effects—at least enough to require that individual nuclear attacks be mounted against each satellite, if our satellites are adequately dispersed. Looking far enough into the future, it is possible to postulate a widely deployed system of very powerful pulsed lasers or other energy beam weapons in the Soviet Union, each capable of destroying a satellite with a single attack. If that sort of capability ever becomes a strong possibility, it may become necessary to offset it by applying "stealth" technology to the High Frontier satellites, and deploy large numbers of small decoys. This would make an attack against the defensive system drastically more difficult.

There are important military responses to the wartime Soviet attack modes. All attack modes that depend on launch of a Soviet missile or satellite into space would be subject to attack by any space defensive system which was also capable of performing its primary mission of attacking ballistic missiles. The Soviets could no longer count on mounting a successful attack against passive space systems. Further, all chances of destroying the U.S. deterrent on the ground would be lost.

We conclude that the survivability of the High Frontier defenses would give them unprecedented strategic value. The following pages outline some of the possible consequences of the U.S. pursuing a High Frontier defense. We believe it is mandatory that we reduce the military risks we face—hence funding of ballistic missile *defenses* is a true imperative. Such funding is clearly of higher priority than funding any other kind of incremental increase in military capability.

Our principal alternatives in reacting to a Soviet first strike range from choosing to accept such a strike passively (absorbing the consequences), to launching on warning, to actively defending against the attack. In this brief discussion, active defense will be delineated as point defense; GBMD-1 defense; or GBMD-2 defense.

For example, point defense means close-in defense of Minuteman silos only. Radar would be employed to trigger the reentry vehicle (RV) kill mechanisms— intelligently guided rocket interceptors, swarms of small projectiles, or possibly high-fire-rate guns. Since they are nonnuclear and of short range, they could be autonomous systems; they would fire automatically at any very high-speed object coming in on a particular course. Kills could be effective at ranges as close as three to four thousand feet. Thus, decoys would not be a problem (they would have burned during reentry).

Our first-generation spacebased defense, GBMD-1, would consist of orbital platforms, each with 30 to 150 guided interceptor projectiles, continuously circling the Earth in orbits that would place them over the desired Soviet target areas for only a fraction of each orbit—hence many platforms would be required. Between 200 and 500 are contemplated. Therefore, several thousand interceptor projectiles would be necessary in order to have about 1,000 always usefully close to the Soviet launch complexes. Boost phase interceptors would use readily available infrared sensors to home in on the heat of an ICBM's propulsion units.

Since GBMD-1 would have hundreds of extra interceptors which would come into useful ranges while RVs are in midcourse, these interceptors would be available to kill RVs that elude the boost phase interdiction effort. Midcourse intercept, however, may require the development of second-generation hardware, GBMD-2, with improved sensors to see the small and now cooler RVs to provide adequate intercept guidance.

If an international situation should develop in which the Soviets believed that the gains to be gotten would justify a surprise attack, what would be the consequences to the U.S. if we chose *not* to have active missile defenses in the future? Assuming an 80% net kill probability against Minuteman, the Soviets need target only two RVs per Minuteman silo to achieve a 96% annihilation of our Minuteman force. Only about

40 of our 1,000+ Minuteman force would survive. A parallel attack on 1,000 other softer targets of military value would result in only 1%, or 10 of these, surviving.

If the Soviets view a U.S. launch on warning option as a credible policy, it could be a strong psychological deterrent. It could buy us valuable time while we build a ballistic missile defense. However, in the event of a Soviet first strike, it is nonetheless a hollow policy. If we succeed in launching on warning (obtaining launch authority from all levels concerned in a very short time—even if Soviet ICBM launch is not positively verified), then there will be a massive exchange. Otherwise, we absorb a unilateral first strike as previously discussed. Whether we absorbed the strike or launched on warning, almost all of our landbased military resources would be destroyed by the incoming Soviet attack.

A groundbased point defense would in itself yield very different consequences. Using existing technology and off-the-shelf components, we could rapidly obtain point defense of 20% of Minuteman. If our system employed simple swarms of small projectiles, the kill probability of the system might be as low as 0.4. Nonetheless, given this limited deployment and low effectiveness, the survivability of Minuteman would be *more than double* that of the previous case—86, versus only 40, surviving. If we adopt a more effective system, e.g., a multiple-shot system with high kill probability, then Minuteman survivability against Soviet attack might be over 70% (see Figure 3).

Figure 4 illustrates the value of a GBMD-1 spacebased missile defense system. Even if such a system exhibited kill probabilities of only 50%, the system would assure that over one-third of the Minuteman

MINUTEMAN SURVIVING

		Percent MM Protected	
		20%	100%
Defense Kill Probability	.40	86	270
	.80	174	706

Other Targets   10

Figure 3. Ground-Based Point Defense

force would survive. Actually, preliminary engineering studies suggest that kill probabilities could be over 90%. Consequently, Minuteman survivability is shown when protected by a GBMD with a somewhat more conservative 85% kill effectiveness. The survivability of soft targets may be fully as important as having our missile retaliatory forces survive, because it is most likely that a Soviet ICBM attack would be coordinated with ground forces moving toward preselected objectives.

	Surviving Minuteman	Other Military Targets
P_{KILL} = 0.5	360	302
P_{KILL} = 0.85	774	748

Figure 4. GBMD-1 Interdiction

In Figures 5 and 6, the effectiveness of defense in depth is estimated. With midcourse GBMD-2 interceptors taking their toll of any ICBM or RVs escaping the first defensive line, Minuteman survivability jumps to over 90%. The triple-layer defense (Figure 6) includes GBMD-1 (boost), GBMD-2 (midcourse), and point defense for Minuteman. The cumulative effectiveness of such a defense should totally deter nuclear attack via ICBM, either as a surprise or as the consequence of escalation to theater level (or higher) warfare.

	Surviving Minuteman	Other Military Targets
$P_K = 0.5^*$	640	600
$P_K = 0.85^*$	964	960

*Both boost phase and mid-course interdiction

Figure 5. GBMD-1 Plus GBMD-2 Interdiction

	Surviving Minuteman	Other Military Targets
Low P_K	774	600
High P_K	993	960

Figure 6. Point Defense Plus Space Defenses

Spacebased defense provides survivability to both our retaliatory missile forces *and* our conventional military forces (see Figure 7). The military value of the survivability of these forces is incalculably high.

First, the likelihood that we would survive a surprise nuclear attack with most of our missiles *and* our conventional forces intact should preclude any Soviet attempt to make such an attack. Therefore, these defenses constitute a *positive* deterrence to possible Soviet opportunism—a deterrence not based on Western intellectual presumptions about the unacceptability of nuclear warfare to the Soviet Politburo or military leaders.

	Units Existing (1981)	Surviving Units (High P_K's)				
		MX-MPS (200 MX) (No Defenses)	100% Point Defense	Boost Phase Defense	Boost & Mid-Course Defense	3-Tier Defense
	Ref.	1	2	3	4	5
ICBM's	1000	45	706	774	964	993
Air Bases	140	1	1	105	134	134
Army Bases	60	1	1	45	58	58
Naval Bases	70	1	1	52	67	67
Overseas Bases	140	1	1	105	134	134
Other (Nuc., Fuel. . .)	590	6	6	441	567	567
Total	2000	55	716	1522	1924	1953

Figure 7. Survivability—All Military Targets

Second, in the event that limited or widespread nuclear war does ensue, we will have maximum chances of using our conventional military forces during and after any exchanges. Since nuclear warfare is the worst possible deterioration of military confrontation, it would be undertaken only if an extremely valuable objective were to be gained. Therefore it is presumed that the exchange must be followed by conventional force moves to capture the desired objective. Thus the retention of maximum general-purpose military capabilities *after* nuclear attack would appear to be an imperative of the highest order.

Finally, the existence of active defenses creates uncertainty—uncertainty that a Soviet strategic planner cannot resolve. The military value of such uncertainty could be pivotal, because this factor could preclude the Soviets from even considering many options that they may otherwise find to their advantage.

CHAPTER IV
THE NONMILITARY DIMENSION

Space holds the prospect for a return to sanity in the national security sphere. It also holds the prospect for a revitalization of progress beyond the military sphere. With a proper combination of space technologies, we can sharply improve the security of the U.S. and its Free World allies, and can also restore confidence in the ability of Free World economies to meet the challenges of the future.

Few scientific or technical events in human history have been as truly revolutionary as the conquest of space. Man is still too close to his recent entry into the space age to appreciate fully the impact of our evolution into space and its implications for economic growth, political influence, and national security. The extraterrestrial expansion of the human species will affect everyone. Already, the effects on communications and Earth observations are global. The conquest of space and its resources will be as important to economic expansion in the 21st century as improved ships, navigation techniques, and firearms were in the development of resources in the 19th century.

During the last 150 years, the industrial revolution affected lifestyles in every part of the world. New industries profoundly changed the relationships between nations. The 19th-century empires disintegrated as the new political and economic dominance of individual nations, largely based on their technological progress, led to the East-West struggle for global influence and the emerging of the North-South dialogue. Technological progress will continue to be the key to economic growth, political influence, industrial expansion, and national security.

Space is the high frontier that will be recognized as having the strongest influence on future strategies, both commercial and military. The success of these strategies will have profound effects on the resolution of many contemporary concerns, ranging from the availability of assured energy resources to meeting Third World economic hopes.

In the past, government institutions and industrial organizations usually concentrated their planning and decision-making on the near term, often less than five years. Space-utilization strategies require long-term planning, extending over the next several decades. Such long-term commitments of resources imply a distant horizon for space programs. This fact may adversely affect decisions on development funding for space enterprise—yet it is the most promising option of the 21st century. For example, space missions such as the manned lunar landing of Skylab achieved their short-term objectives, but failed to provide any visible forward thrust for the next evolutionary steps. What we need is the development of core technologies which can meet the requirements of *many* space applications, both near and long term; technologies that can thereby maximize both the commercial and military uses of space. If we abdicate our hard-won lead in space, the result could be the rise — once again—of imperialism, using modern technology to the same ends as imperial powers did in the last century.

A key factor in the success of future commercial space programs could be international cooperation among Free World nations. This would require an international framework capable of coordinating, integrating, and managing the efforts of contributors in many parts of the world. With such cooperation, it is likely that the potential industrial uses of space can meet our expectations without placing an undue burden on U.S. industry or taxpayers.

Industry, which must focus on the risk as well as the profit potential of ventures in space, will need government support. Industry cannot gamble with investor funds. Therefore, when high-risk, long-term development periods or large capital requirements become necessary, joint industry/government cooperation will be essential.

In brief, the U.S. must be prepared to meet both the military *and* the economic challenges of space. If successful, the U.S. will gain much more than the prestige of a Sputnik triumph or a Moon-walk drama. The U.S. will garner the priceless advantage of security for its citizens, and the ability to supply valuable new services, products, and energy to the world.

This last-cited opportunity, energy from space, has an especially promising potential. A decade of scientific studies and a recent comprehensive assessment—societal, economic, environmental, and technical—by the Department of Energy conclude that power can be beamed from space to Earth in quantities so large as to supplant traditional power plants. For example, it was estimated that *one* solar power satellite (SPS) could produce the electrical power of up to *five* nuclear or coal plants. Further, the cost of this power would be competitive with terrestrial power options. If SPSs are successfully developed, they can supply electrical power on a global scale. Our economic gains and our independence from nonrenewable energy resources would be extremely valuable strategic assets.

It is time for the United States to embark on a carefully planned commitment to exploit space. A piecemeal, on-again off-again approach, typical of our past efforts, will not prove timely or economical. In contrast, Japan establishes long-term goals and stays with them until they are successful. European consortiums are also increasingly setting long-term goals. Where will we be, one or two decades hence, if we do less?

History is likely to recognize only two U.S. space efforts as crucial—our national commitment in the 1960s to be first on the Moon, and our national commitment now to develop the strategic potential of space, if indeed we do seize this opportunity. If we do not, we may be denied security in space, access to spacebased energy, products, and services. We would then be denied recognition as a leader in technology. Other denials would closely follow. This Administration is at the right place, and this is the right time, to claim an historic opportunity for the United States.

How do we exploit this opportunity? Initially we need to develop a more economical space transportation system and to fund the low-cost, preparatory and developmental stages of the most promising space industry opportunities. Only after the means and economic viability of commercial ventures are established need we make large commitments of government or private funds.

The concurrent development of civil and military space support systems will produce valuable synergistic benefits to both civil and military programs, because both depend on the same core technologies. In the past, the U.S. has erected artificial barriers between the two efforts. These barriers have been grossly detrimental to both space programs. However, under the present Administration, DOD and NASA are making renewed efforts to break down those barriers.

Another aspect of joint development which we must not overlook is major investment in commercial space systems. We cannot expect individual nations, companies, or consortiums to create the space facilities necessary for industrial/commercial growth if those investments are not protected from hostile interference, attack, or seizure.

Our proposed space policy's most important initial objective is the development of an improved Space Shuttle. Our first goal should be a substantial payload increase to garner the benefits associated with the economy of scale, as well as a desirable increase in unit capability. Our second goal should be system cost reduction. The total achievable cost reduction, using current technology, was conservatively predicted in recent engineering studies to be ten-to-one when compared with the current Shuttle. A recent technical assessment by NASA (I. Bekey and John E. Naugle, *Just Over The Horizon in Space, Astronautics & Aeronautics*, NASA Headquarters, May, 1980) projects the following potential cost reduction: "The Shuttle will not do better than $1,000 to transport one kilogram to orbit, compared to only $5 to fly one kilogram in an airliner from Los Angeles to New York, although the energy requirements are the same. The cost of the equivalent electrical energy comes to only about 50 cents, leaving a lot of room for improvement." The NASA authors expect fully reusable vehicles and other systems to reduce the cost of space transportation " . . . *by at least two orders of magnitude.*" An optimized system would operate like an air-cargo airline. It would operate on a rapid turnaround schedule assuring high utilization of ground-support facilities and people.

The second immediate objective of our new space policy should be the creation of a manned space station in low-Earth orbit. Aside from military systems, such a station in near space would permit low-cost, efficient development and testing of system elements of commercial interest. The later deployment of a manned station at geosynchronous orbit (some 22,300 miles from Earth's surface) is also a sound economic investment. The first station, then, may be designed as both an initial test facility and as a "way-station" for transition to a sustainable, manned presence in geosynchronous orbit. Eventually, the Moon itself may serve as a space station with important benefits for the construction of installations, such as SPS, in geosynchronous orbit.

Our third immediate objective should be the development of reliable, high-capacity energy systems, such as SPS, in space. The initial application of an SPS would be the powering of other space installations. The success of an SPS would almost certainly engender public confidence in commercial space investment.

The final objective of our proposed space policy is the preparatory development of a selected number of promising commercial business opportunities. The term "preparatory development" was adopted to embrace the following considerations: the *ultimate goal* of all this effort is to establish independent, self-reliant space businesses, and any government efforts to "seed" these ventures should be oriented toward this goal. The government's efforts should focus on preparing for the transition of these seed efforts into *independently viable* commercial operations as soon as possible.

These objectives imply that system developments should emphasize production, rather than research goals. We must not underestimate the importance of this emphasis. Research-oriented programs seek to glean as much knowledge as possible from each development. However, undesirable effects usually occur, including increased development costs and program delays. In order to ensure that we prevent such effects, we should have oversight boards—with board members from business—to preside over any preparatory development that is destined to be a commercial business. Means of meeting the first two objectives (improved space transportation and manned space stations) are being advanced by the Boeing Company, Martin Marietta Corporation, Rockwell International, and others. Several proposals also exist for powering space installations: nuclear power plants in orbit; reactors on the Moon; and several options for the conversion of solar energy into electrical power. These systems require more study before specific recommendations can be made on preferred options. We can state, however, that the strategic potential of energy from space is so enormous that vigorous research on these options is essential.

SPACE TRANSPORTATION AND SPACE STATIONS

Today's Space Shuttle, a returnable space transport, is proving more economical than any previous expendable rocket system. However, it was designed to use as much available hardware as possible (its strap-on boosters, for example), because its *development cost* had to be minimized at the time it was approved. This was due to NASA funding limitations. Consequently, its *economic* performance is less than technology would have permitted, even at the time it was designed. If we now develop a larger carrier, and pay careful attention to ground cost minimization, engineering studies predict that that future space transport costs could be cut to ten percent or less of current Shuttle costs. Such a cost reduction would profoundly increase the capability per dollar spent, and would greatly expand the range of business opportunities that will prove to be viable, because commercial costs are very sensitive to space transportation expense. Clearly, the development of a more economical Shuttle is the highest priority item for both military and commercial programs outlined in this book.

A second-generation Shuttle would most likely be a two-stage vehicle with both stages fully reusable (see Figure 8). Routine reuse would be patterned after cargo airline operations; so fast turnaround, high maintainability, and minimum life-cycle cost will be primary design requirements. First-stage booster fuel would be an inexpensive hydrocarbon, e.g., methane. This stage could be of all-aluminum construction—heat sink design rather than heat shield tiles—since the first stage separates and returns from the fringes of the atmosphere at less than orbital velocity. The pacing development item would be the main booster engines. Based on system-design studies performed for NASA, a 125-ton payload would require five or six engines of approximately Saturn F-1 size and type, but redesigned to provide longer life, low cost, and easy maintainability. A development cycle of only five years is possible with responsive program management. Total program cost would be in the vicinity of $12 billion.

The initial space station at low-Earth orbit could be the space operations center (SOC) now advocated by NASA. This platform would be an operational base in space for the assembly and test of space equipment, repair of satellites, and the staging of equipment bound for higher orbits. Minimal scientific research would be conducted here (see Figure 9). With a 1983 go-ahead, a space operations center could be in place by 1987. Costs are estimated at approximately $6–8 billion.

SPACEBASED ENERGY SYSTEMS

The Solar Power Satellite (SPS) would employ some proven hardware, since solar cell arrays in orbit have already been used to collect energy from the Sun. This power can be relayed to Earth through microwave transmission. The microwave transmission system requires large antennas in orbit and on Earth to be efficient. For example, in one design a 3,000–foot-diameter transmitting antenna is part of the SPS, and a four-by-five-mile elliptical receiving antenna is required on the ground (see Figure 10). If a high level of power, say 5,000 megawatts(!), is transmitted from orbit to Earth, then the unit cost of energy can be very low.

Figure 8. An Advanced Space "Shuttle" Design

Because of large-scale operation of the system, delivered power costs of an SPS are predicted to be competitive with coal or nuclear power plants. For example, if a $12.5 billion ($2,500 per kilowatt in 1981 dollars) system capable of 5,000 megawatt output were purchased, over a 40-year period it might cost around $78 billion to own and operate it; $12 billion in depreciation plus $21 billion interest at 12%, $33 billion earnings at 18%, plus another $12 billion in operating expenses, taxes, and other costs. The station would deliver 1.6 *trillion* kilowatt hours of power over that 40 years. Hence, the *average* cost of the power is under five cents per kilowatt hour.

A comprehensive assessment of a representative spacebased energy system was conducted by the Department of Energy from 1977 to 1981. Their evaluation did not reveal any technological barriers. The next step is to continue system definition, refinement of initial socio-political-environmental assessments, and tests of key system elements. Since cost attainment is vital, one major program emphasis would be cost control. Finally, cost attainability would have to be demonstrated *prior* to seeking funds for full-scale implementation. A conventionally paced research program would require spending about $30 million per year on energy systems for the next three to five years. At the end of this period, program leaders would seek commitment to pilot production of key items (to demonstrate cost achievability), and a limited demonstration of promising technologies in space. The first full-scale energy system in space would be built after 1995 if the system's promise is achieved.

An Administration decision to support a joint commercial/military space program has many political

Figure 9. A Space Station

Figure 10. Solar Power Satellite

ramifications. We believe that the majority are strongly positive. The dominant benefit to the U.S., in embarking on such a program, is that this country will then have a comprehensive space policy with well-integrated *long-term* objectives. If, in addition, the U.S. offers participation in these space efforts to our allies, many international benefits can accrue. For example, the proposed High Frontier space defense is capable of protecting Europe and Japan, and our allies would share in developing spacebased industry; so they would be likely to share in the costs of the program.

CORE TECHNOLOGIES

We must orient our near-term space programs so that they provide information on which to base the next phases. Thus, our near-term programs must feature the development of *core technologies* which can meet the requirements of many space applications, and which will lead to step-by-step advances in the commercial and military uses of space (Figure 11). As important as the role of technology is in the planning and execution of specific space programs, other issues — economic, environmental, and societal — must be considered in parallel to ensure continued broad support for the space programs.

Industry, which must focus on the near-term profit potential of ventures on Earth or in space, will need the stimulation of government support. Without this support, industry will not embark on activities requiring long-term, high-risk investments. Joint industry/government cooperation and planning will be essential to achieve the longer-term commercial space pro-

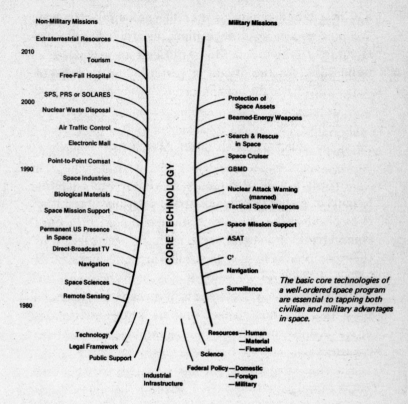

Figure 11. *Core Technology and Space Applications*

gram goals of creating profitable markets for industry.

Inevitably in today's economic climate, space activities are going to be assessed more critically than they were two decades ago. But financial constraints should not be the sole determinant to the planning of industrial activities in space. A close coupling between these activities and national security implications is imperative. Beyond that, the activities should be part of the strategy to find synergistic solutions to humanity's most pressing problems pertaining to energy, the environment, and resources. What we must have today is a broad perspective of the potential of space. We must engage in strategic planning, guided by Robert Goddard's statement: "It is difficult to say what is impossible, for the dream of yesterday is the hope of today and the reality of tomorrow."

INTEGRATED SPACE MISSIONS

There is a wide divergence of views regarding the long-term global impacts of space missions; the influence of advances in science and technology on future commercial, industrial, and military activities; and their competitiveness with similar activities performed on Earth. Nor do we enjoy a close consensus on the scale, timing, and effectiveness of investments in space programs by industrialized nations in response to idealistic visions, pragmatic considerations, and political realities.

The U.S. space program has pursued two separate paths. One path is representative of the nonmilitary activities of NASA and, more recently, by industry — primarily in communications. The other path is followed by DOD, which pursues military activities recognized as critical to U.S. national security. The nonmilitary space missions fall into broad categories of information, energy, industry, services, and science. Military space missions are concerned with intelligence, defense, and offense. Although the budgets of the two programs are about the same, there have been only limited interactions between them. Obviously the programs can be of mutual assistance. This was recognized in the joint NASA/DOD development of the Space Shuttle, and deserves more attention as we undertake more ambitious missions which could use a similar technological base.

In view of the large investments required for future space missions to achieve military and commercial goals, we must develop an integrated long-term strategy based on incremental advances of technologies of increasing scope and effectiveness. This strategy is already pursued by the USSR, where it is sometimes difficult to discern the boundaries between military and nonmilitary space missions because the development of specific technologies is common to both. Our nonmilitary space program, on the other hand, has been characterized by well-publicized missions (e.g., Apollo, Skylab, and Voyager) which were undertaken to meet specific political, technological, or scientific goals. Despite the wealth of information we obtained, these missions resulted in a dead end because we had not evolved an integrated long-term strategy for space applications.

Space missions can be viewed as branches of a tree trunk, fed by core technologies—the strong roots required for further growth (see Figure 11). These core technologies should be developed to support commercial and military missions as well, with each branch providing increased capabilities and using more substantial capital investments. The success of near-term space missions would reduce the risks of later missions, and would also help pinpoint the core technologies which need further attention. In this way, research and development will support the largest number of future mission goals. For example, the goal of achieving a permanent U.S. presence in space would have significant implications for both commercial and military missions. Military space planning today does not include the construction of large space structures, advanced robotics, power plants in space, or manned operations in geosynchronous orbit. We should analyze the commonalities of core technologies serving the goals of both these programs, so that we can use space-program budgets most effectively in support of a national space policy. The evolution of a national space policy, which recognizes the potential of space to meet a variety of national goals, deserves a high priority.

SPACE RESOURCES

The benefits of nonmilitary space missions are closely coupled to the unique resources of space. Space is not merely a medium for passage while observing and supporting activities on Earth; its resources are keys to a broadening array of applications. The primary resources of space include:

- Absence of gravity in satellites in various free-fall orbits.

- Benefits of gravitational forces which help determine shapes of orbits (for example, by attaining geosynchronous orbit some 22,000 miles from Earth, we can place an SPS stationary with respect to a desired location on Earth below).

- Vacuum for industrial processes; a more nearly perfect vacuum than can be obtained on Earth.

- An infinite heat sink only a few degrees above absolute zero, permitting us to dispose of unwanted heat by radiating it into deep space without thermal pollution effects on Earth.

- Electromagnetic radiation—primarily solar energy which awaits our collecting and converting it.

- Presence of magnetic and electrical fields.

- Unobstructed view of Earth from low Earth, Sun-synchronous, elliptical, and geosynchronous orbits.

- Availability of materials from the Moon, other planets, and (especially) the asteroids—materials which could forever remove the limits to growth of human civilization.

The last-cited item involves potentially staggering quantities of some materials; quantities so enormous that Earthly mining output seems minuscule by comparison. Given man's proven ability to work on the Moon and in orbit, we still need supplies of resource materials, e.g., minerals and gases. At present it is still prohibitively expensive to carry such raw industrial materials into space from Earth—fighting the planet's "gravity well" and its atmosphere en route.

But the Moon has minerals, only one-sixth of Earth's surface gravity, and no atmosphere. There are several means by which we could transport mined materials from the Moon's surface into lunar orbit for industrial purposes. Easily the most promising source of raw materials, however (and perhaps the most inexpensive), is the Asteroid Belt, between the orbits of Mars and Jupiter. Indeed, some asteroids follow orbital paths that carry them quite near Earth at times. We have already sent probes far beyond this region. The size of these tiny planetoids varies from the inconsequentially small to some so large that they have names (Ceres, well over 400 miles in diameter, is the largest). Many tens of thousands of these objects are known to exist, and their composition is believed to be rich in both oxygen and minerals, including metals.

Nor would we be obliged to process an asteroid where we found it. The gravity of an asteroid is insignificantly small, and it has no atmosphere. A mile-diameter chunk of rock and metal could be eased from its orbit and into Lunar orbit if we chose. NASA studies reveal that asteroid-capture missions could be devised in which low energy expenditure is the primary consideration. Asteroid mining may be more cost-effective than mining the Lunar surface. We could process an entire asteroid—or, mining it by hollowing it out, obtain a secure habitat that could absorb tremendous external impacts without endangering those who worked and lived within.

The wide range of nonmilitary space missions now being performed, or in various planning stages, is shown in Figure 12. The broad mission categories include information acquisition and dissemination; energy supplies; industrial activities; commercial services; and scientific investigations. They can also be classified according to orbits, e.g., low-Earth (LEO) and geosynchronous (GEO); lunar location; or missions which encompass the solar system. The striking features of current nonmilitary space missions are their global impact and the growing level of investments in information services and communications. These missions yield core technologies for future industrial and commercial activities. They can therefore create profitable opportunities for private investment, and can lay the foundation for a dramatic expansion of space activities.

Location	Information	Energy	Industry	Services	Science
LEO	● Remote sensing ● Mineral resources ● Agriculture ● Fisheries ● Search and rescue ● Disaster relief ● Earthquake prediction ● Hydrology ● Fire detection ● Pollution monitoring ● Border surveillance ● Cartography	● Sun-synchronous SPS ● SOLARES	● Biological materials ● Solid-state devices ● Advanced alloys ● Improved magnets ● Optical components ● Superconductors ● Freefall casting ● Space construction	● Space mission support ● Tourism ● Freefall hospital	● Optical astronomy ● Infrared and ultraviolet astronomy ● X-ray astronomy ● Materials science ● Biology ● Solar physics ● Geophysics ● Oceanography
GEO	● Communications ● Direct-broadcast TV ● Point-to-point communications ● Electronic mail ● Education ● Weather forecasting ● Navigation ● Search and rescue	● Photovoltaic SPS ● Thermal conversion SPS ● Fast breeder reactor ● Power satellite ● Power relay satellite	● Hazardous materials ● Space construction	● Space mission support ● Air traffic control ● Night illumination ● Biological isolation	● Radio astronomy ● Geophysics ● Solar wind studies ● Meteorology
Moon		● Solar power breeder reactor	● Mineral resources		● Radio astronomy—(farside) ● Planetology ● Geology
Solar system		● Nuclear waste disposal	● Asteroid resources		● Planetology ● Geology

Figure 12. Nonmilitary Space Missions

Activity	Information	Energy	Materials	People
Major space advantage	● View Access	● Solar flux	● Low g ● High vacuum	● Uniqueness
Major technical	● Antenna size—10m to 100m ● Power—21 to 10,000 kW ● Data processing	● Power output—1 to 5 GW ● System mass—10^4 tons ● Cost—$\$10^{10}$ ● Transport cost—$<\$20$/lb LEO ● Environmental effects ● Societal issues	● Proof of theory ● Production and process development ● Power—10 to 10,000 kW ● Transport cost—$<\$100$/lb LEO	● Transport cost $<\$25$/lb ● Habitation
Timing for significant revenues	● Present $>\$300$ M/yr	● 2010 +	1990 +	● 2010 +

Figure 13. Space Industrialization Activities

The industrial uses of space, in addition to those we have cited, include processes to produce unique materials by taking advantage of the space environment; disposal of hazardous materials by selecting trajectories so that they would be consumed by the sun; space construction to support industry; and people-oriented services. Figure 13 lists industrial activities which could take advantage of space resources. It also indicates the major hurdles which we must overcome to achieve goals related to information, energy, materials, and people. For example, enhanced data processing and information-exchange capabilities would require increased antenna sizes and power supplies. Solar energy could lead, as we have said, to SPSs which would beam power both to other satellites and to Earth. Large mirrors could reflect solar radiation to desired areas on Earth so that terrestrial solar energy-conversion facilities could be used 24 hours a day.

Growth Potential

Efforts are already underway to confirm the possibility of producing unique materials in space, e.g., alloys which cannot be produced in Earth's gravity. Once we have demonstrated that we can produce such materials, and have information about their structures and properties, we can develop production methods and processes with a view toward establishing space industries. The low gravity conditions in orbit, possible attractive features of living in space, and utilitarian motives associated with space industry, may result in a gradual increase in human habitations in space. Although it is too early to project whether large-scale human migration will take place, the demonstration of habitability without adverse effects can be expected to result in a permanent human presence in space.

The time for the growth of space industry extends well into the 21st century, with significant revenues projected for each of these activities. The potential for profit will, in large measure, depend on successful development of economical, reliable space transportation. The thrust of space transportation system development is exemplified by the Shuttle, whose goal was to significantly reduce transportation costs to low-Earth orbit. There is no inherent technical barrier to development of a transportation system which could approach airline procedures in support of space industry.

In a little over two decades, mankind has penetrated outer space, brought minerals from the Moon, and obtained highly beneficial global information from satellites, e.g., weather information. It is inconceivable that our evolution into space will not grow exponentially as space industry opens new markets, demonstrates expanding business opportunities, and becomes the arena for national cooperation as well as competition. Clearly, we must meet many political, social, legal, and financial challenges on the national and international scale, so that the tangible returns will be of the widest benefit to society while rewarding those who are creating new industries.

The projections of markets and potential revenues are, naturally, based on necessary assumptions about future space industry. However, the markets for information and energy-related activities are already so great (see Figure 14) that, even if annual revenue falls short of the projected increase rate, the revenues of space industries could *still* be among the fastest-growing of all industries in the 21st century. The industrialization of space could have as profound an effect on nations of the next century as the industrial revolution had on political realignments of the 19th century.

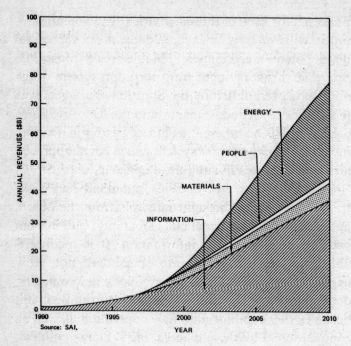

Figure 14. Projected Space Industries Revenue Comparison

PARTICIPATING NATIONS

The growth of space industry is not restricted to only a few nations; it includes an increasing number of participants (see Figure 15). The Soviet Union has demonstrated its commitment to a vigorous long-range expansion of its space capabilities, with emphasis on manned space operations lasting many months. These are the precursors of *their* space industry, in parallel with their continuing commitments to substantial military space operations. Several European countries are increasing their space-related activities, and Third World countries increasingly recognize their stake in the successful outcome of space industrialization. One of several striking examples of growing commitment to space programs is the Japanese space industry's projection that their annual sales will grow tenfold, from one hundred billion yen annually in 1980 to one trillion yen in the mid-1990s. The Japanese expect to share not only in the growth and demand for various satellites, but also in the manufacture of products in space and in the development and launch of rockets. The Japanese realize the value of a high-technology program as a stimulus to their economy and as a means to improve the quality of life in Japan. The Japanese report states: "The world is in a major transitional stage to enter the space utilization and Japan, too, should ride on the wave of the future and move steadily forward—it is necessary to plan for Japan's development of the space industry in line with the U.S. and Europe from the standpoint of a long-range view and global outlook."

Another example of international interest in future space activities is demonstrated in the studies of the SPS concept. International meetings of professional societies reveal that SPS studies have been made by the U.S., Canada, England, France, Germany, Japan, and the Soviet Union. In addition, Austria and India have established industry working groups under government auspices to follow progress in the SPS field.

Already a number of organizations have been founded with the aim of advancing space industry (see Figure 16). The largest and most successful international organization supported by governments is INTELSAT. Most likely, these organizations—international, governmental, and commercial—will proliferate as transportation costs drop, orbital operations rise, and

Nation	General						Unmanned				Manned		Unmanned					Manned	
Capability	Data handling	Ground station(s)	Launch facility(s)	Suborbital launch	LEO launch	LEO return	GSO launch	Orbital rendezvous	LEO launch	LEO occupancy	Orbital rendezvous	Orbital propellant transport	Earth observation	Communications	Navigation	Test and experiment	Maintenance and repair	Remote control	Materials processing
United States	X	X	X	X	X	X	X	X	X	X	X		X	X	X	X	X	X	X
USSR	X	X	X	X	X	X	X	⊗	⊗	⊗	⊗	⊗	X	X	X	⊗	⊗	X	X
China (PR)	X	X	X	X	X	X							X	X					
France	X	X	X	X	X	?							X	X					X
India	X	X	X	X	(X)								X	X					
Japan	X	X	X	X	X		X	X						X					
ESA	X	X	X	X	X		(X)			(X)			X	X			(X)		X
Other	X	X	X	X	X								X	X				X	X
Total number of nations	111	39	24	15	9	3	4	2	2	3	2	1	7	13	2	3	2	3	5

(X) To be demonstrated.
X Capability has been demonstrated.
X Currently unique capability.
Source: SAI.

Figure 15. Global Space Industrialization Capabilities

	Formation date	Organization	Composition	Activities
Governmental (International)	1964	● Intelsat	104 nations	● Intelsat series
	1971	● Intersputnik	9 nations	● Molniya
	1975	● European Space Agency (ESA)	10 nations	● Many (including Ariane as shuttle competitor)
	1975	● Nordic Telescope Satellite Committee (NTSC)	3 nations	● Nordsat (Ariane-launched)
	1976	● Arab Satellite Telescope Organization (ASTO)	20 nations	● Arabsat
	?	● International Maritime Satellite System (INMAR SAT)	40 nations	● Not yet ratified
Commercial — Multinational	1961	● Eurospace	96 companies; 31 banks	● Promotes Eurospace independence (especially launch vehicle)
	1971	● Cons. Ind. Fr-Al. Symphonie (CIFAS)	4 French companies, 4 FRG companies	● Symphonie series
	1976	● MESH	5 European companies	● OTS/ECS
	1977	● Eurosatellite	3 European companies	● H-SAT
	1977	● ARCOMSAT	2 French companies, 1 FRG company	● Proposing Arabsat
Commercial — National	1975	● Orbital Transport and Rocket Co. (OTRAG)	FRG inv.	● Private launch vehicle (Libya)
	1976	● Spacelab Utilization Working Group (ANS)	3 German companies	● Spacelab exploitation

Source: SAI.

Figure 16. Organizations Aimed at Space Industrialization

supporting core technologies expand. Among the most important technology developments is advanced robotics in Japan, the U.S., and elsewhere, which may make it possible to reduce the requirement for labor-intensive activities in space because we can place increasing reliance on automated construction and other industrial space operations.

SPACE LAW

We must give serious attention to the legal aspects of the use of space. In particular, we must make certain that U.S. industry is not impeded unnecessarily in its exploitation of commercial space opportunities by ill-conceived international legal systems. Highly idealistic urges to preserve the use of space for "all mankind" have already resulted in U.S. acquiescence in the creating of a body of international space "law" —a set of constraints that we have since recognized as detrimental to U.S. economic interests. If we do not reverse this trend, such high-mindedness may result in the *denial* to mankind of the benefits of space industry.

GOALS OF SPACE INDUSTRIALIZATION

The broad goals of space industrialization are to provide means to meet the most vexing challenges found on Earth: the dwindling of natural resources; environmental degradation; and the hopes of humanity to achieve a better standard of living. Space industry is synergistic with national security. Extensive space operations will bolster the leadership of U.S. high-technology industry, and will also provide the industrial infrastructure required for both commercial and military space operations. We are unlikely to build large space-information systems, materials-processing plants, and SPSs by an Apollo-type national effort. Rather, these things will be achieved by a vigorous national space program which recognizes that a series of well-defined evolutionary steps, solidly founded on supporting core technologies, will be required (see Figure 17).

The year 2000 has been chosen as an arbitrary dividing line for those space applications and core technologies. We find a reasonable consensus that they could be developed during the next 20 years. The investments indicated in Figure 18 assume an average funding level of about $5 billion per year. The revenues during the first 20 years of industrial space applications may achieve only 20% to 50% of investments, but they will set the stage for more extensive applications as indicated in Figure 17, when revenues can be expected to rise above investments. We can expect space industry to expand enormously after 2010, when extraterrestrial materials can bring about a new era in human activities in space. We may then have solutions (SPS, for example) to problems of Earthly energy supply, irreversible environmental effects, and depletion of natural resources.

The concept of space industrialization implicitly assumes that industry will make practical use of research and technical developments that have been publicly funded. The challenge to a national space policy will be to reinforce and stimulate growing markets for products and services of space industry, rather than substitute government commercialization activities. If we carry government activities too far, or if any agency remains involved too long beyond the R&D stage, we risk discouraging private investment.

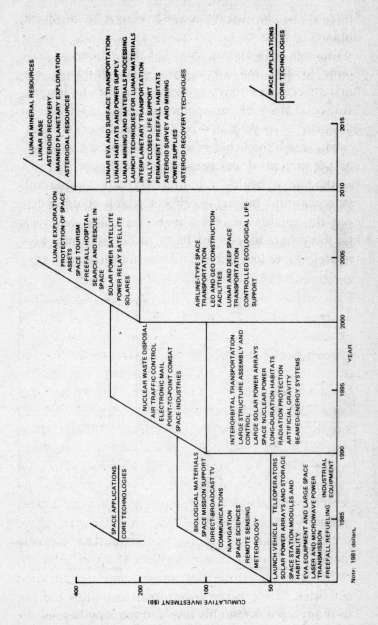

Figure 17. *Nonmilitary Space Application*

Thus, we would also blunt the leading edge of space industry.

Although it is appropriate that we emphasize economic justification for a given space industry, we must also consider environmental and societal issues. The social costs of environmental impacts—whether on Earth or beyond it—must be assessed so that we can weigh the benefits of specific space endeavors against potential dangers. These dangers include risks to human health, potential destruction of natural resources, and intangible effects which could influence the quality of life. The benefits and costs of specific programs are not likely to be uniformly distributed. More likely, they will be concentrated in cer-

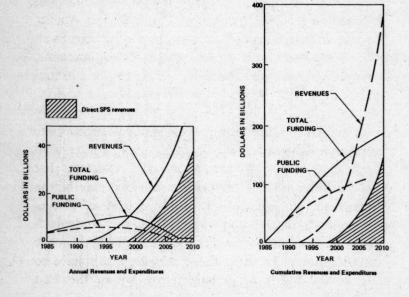

Figure 18. Space Industrialization Program Scenario—Cost and Revenues

tain segments of society and the economies of industrialized nations. Individuals, corporations, institutions, and sectors of industry will react to these specific costs and benefits as they perceive them. As a result of varied perceptions, political pressures may arise which could have a profound effect on the direction of a specific development program's schedule and its ultimate success, both in terms of public acceptance and return on investments.

We can expect expansion of space industry to heighten public concerns on such issues as centralization of control of a specific program, and public involvement in program review. To alleviate public concern and to enhance public acceptance, we should provide information on goals, costs, and benefits of our space industries, particularly if we want international participation in these industries. Unlike Project Apollo, space industrialization focuses on many objectives. Each objective is to be achieved in a specified timeframe, requiring predictable investments and resulting in concrete benefits.

Thus, the "one small step for man, one giant leap for mankind" taken in July, 1969, is not an appropriate analogy for many activities which must be integrated over extended periods to evolve the advances that space industry implies. A better analogy may be the stepwise technological advances we achieved during the industrial revolution of the 19th century, and which laid the foundation for technological advances in the 20th century. In turn, these advances will have to serve the needs of global civilization in the 21st century.

The U.S. is still the leader in space technology, although we may be losing our leadership in other fields. The U.S. may no longer be the dominant innovator in nuclear energy, due in large measure to political protests raised against further developments in nuclear power. These protests have created a potentially dangerous antitechnology atmosphere in America. The present Administration is trying to halt further deterioration of nuclear energy programs, but a good deal of time has already been lost. Other countries, including the USSR, are now developing nuclear technologies rejected by the U.S. In other categories of industry, it is obvious that we can expect more public support and more profit for private industry. The field of audio-visual communications has already demonstrated this in space; the demand for private TV satellite relays is growing steadily. These ventures would not have been possible without the initial government space programs. Government/industry cooperation has also proven profitable in the development of pharmaceuticals. The possibilities for other commercial ventures in space should be thoroughly analyzed by the Department of Commerce to determine the full range of feasible industrial programs. Private-sector industrial and financial experts would be invited to participate in this project.

The High Frontier concept also encourages cooperation and cost sharing between the U.S. and the less developed nations. There are, of course, general pros and cons involved in international cooperation, even among like-minded groups of nations. These issues should be analyzed in depth by the National Security Council. One key general issue in international cooperation is that of technology transfer control, which is obviously complicated by the participation of other nations in our space projects. We must obtain firm guarantees from nations wishing to share our space technologies to prevent the self-defeating handover of these technologies to the Soviet Union or other potentially hostile powers (see Chapter V for further discussion).

Summary

In summary, the goals of space industrialization include the following:

- Adoption of a long-range view and global outlook.
- Integration with national space policy planning.
- Consideration of national and international market forces.
- Cooperation between industry and government to achieve joint goals.
- Evolution of institutional structures which acknowledge the needs of the public to benefit from space industry; a legal and regulatory framework so that space industry can achieve its planned benefits; and mechanisms which will allow both public and private investments.

We must have a national space policy which recognizes the inescapable realities of future space applications, so that we can maintain U.S. technical leadership in space. We must see that this leadership can be maintained and translated into practical industrial terms. Only then can our space endeavors strengthen the U.S. economy, develop exportable products and services, create new jobs, and demonstrate that our national agenda for space will serve both national and global goals so that the 21st century opens the promise of the inexhaustible resources of space.

CHAPTER V
TECHNOLOGY PROTECTION

We have said this before: it is absolutely imperative that the United States maintain its technological advantage over the Soviet Union. The degree to which the U.S. can afford to involve other nations, multinational corporations—even U.S. industry—in High Frontier programs depends on assurances that critical, space-associated technologies will not be transferred to the USSR. If the U.S. and its Free World allies are to ensure that the security and economic benefits of space are free from interference of hostile political systems, we must improve safeguards for our core technologies.

THE SOVIET REACTION

From an historical perspective, it is clear that any shift in the strategic balance in favor of one superpower will provide sufficient incentive for the other superpower to exploit new technology to redress the emerging imbalance. The Soviets have demonstrated great determination and effort to achieve military superiority over the U.S. They are fully aware of the consequences due to changes in the political/strategic environment. It should be evident that they are not going to stand idle and passively observe a United States effort to change the strategic equation.

A review of Soviet assessments of the Reagan Administration's national security policy clearly reveals a Kremlin understanding that the new government in Washington is systematically undertaking to redress the shift in the "correlation of world forces". It is also clear that Moscow views developments in the military arena as the central issue. The Kremlin categorically insists that the USSR can and will do " . . . everything necessary . . . " to prevent a shift in the present military balance. It confidently asserts that it can succeed. Speaking in Kiev on May 9, 1981, Brezhnev said, " . . . if we are compelled, we will find a quick and effective response to any challenge by belligerent imperialism."

Marshal Viktor G. Kulikov, the Commander in Chief of Warsaw Pact forces, stated the same idea in the following manner: "The Soviet people . . . will not allow potential enemies to be able to gain the upper hand in any kind of armament or technology."

Considering proven Soviet policies and methods, we can surely expect that the USSR will launch intensified efforts to obtain from the U.S. those technologies which will assure the leadership in the Kremlin that the U.S. strategic programs will be neutralized. It is an oft-proven fact that the Soviets have used Western technology to build up their military capability. A recent report from the Pentagon states, "Without the transfusion of U.S. technology and equipment, the Soviet Union's capabilities would almost certainly have remained at the 10 to 12 year gap of the 1965 era."

The FBI reports that much of this technology transfer involved covert operations by the Soviets. Included is "theft of proprietary information at considerable cost, illegal transshipments of our technology to Soviet Bloc countries, penetration of computer systems, and compromise of employees." The report, authored by FBI Director William Webster, also charges that Soviet officials and visitors came to the U.S. to attend scientific and professional association symposia only to pilfer information that could be applied to military-strategic and intelligence purposes. Webster adds, "There is a considerable threat where foreign agents either steal technical information or corrupt an employee to steal this data for money."

In a recently issued document, *Soviet Military Power*, our DOD states that during the last decade the industrialized Free World has supplied the Soviets with "billions of dollars" worth of "efficient machine tools, transfer lines, chemical plants, precision instrumentation, and associated technologies." The areas in which the Soviets have made great technical strides through witting or unwitting help from the Free World include directed energy weapons, electronics and computers, explosives, precision welding, advanced composite materials (combining great strength with low weight), space technology, and others.

TECHNOLOGICAL COMPETITION

With respect to national defense, the term "technological competition" refers to the efforts of competing political-economic systems to maintain, or to achieve, superiority in high-technology areas that are important to military systems. This history of competition between the U.S. and the USSR dates to 1943 when the

Soviet Union began its effort to develop an atomic bomb. The unexpected orbiting of a Sputnik by the Soviets in 1957 shocked the U.S., and for the first time focused broad public attention on Soviet technological capabilities and objectives. This event also resulted in a rapid development of our own space science and technology. The competition between the U.S. and the USSR has continued in all phases of space programs and in the development of strategic weapon systems. In this era of unprecedented change, our technological strength is the key to our long-range survival as a nation.

American security, like the American economy, stands on a foundation of technological superiority. We need superiority in defense technology for two reasons. First, because our open society tells our adversaries what we are planning in military technology while their secrecy forces us to provide for many possibilities. Second, in military operations we traditionally depend on superior quality to compensate for inferior numbers.

The U.S. continues to hold a technological lead over the Soviet Union in many critical areas vital to our national security, but that lead has been diminishing. In some very important areas it is gone; in others the Soviets are ahead. Moreover, the technology balance is dynamic. In examining the current technology balance and its dynamics, qualified analysts agree that the USSR expends a very large and determined effort, and that they are inexorably increasing their level of technology relative to ours. In fact, they are seizing the initiative in many important areas, e.g., high-energy laser beam and charged-particle beam weapons, surface-effect vehicles, and antipersonnel pressure weapons.

Technological development is molding future Soviet

strategy. From all indications, the Soviet strategy will continue to center on world dominance, with technology as a key factor. A crucial element in our strategy of deterrence is the maintenance of a margin of military advantage through possession of a number of sophisticated technologies.

CONTROL OF TECHNOLOGY EXPORT

Any thrust toward deployment of advanced technological systems requires parallel efforts to protect against overt and covert leakage of those technologies to potential enemies. We must prevent exports of military and military-related ("dual-purpose") technologies from the U.S. to potential adversaries. We must use all legal methods to protect our technological lead, and not contribute to the military strength of potential adversaries.

Since the end of World War II, the U.S. has relied principally on two laws to control export of certain goods and services to all destinations—including our principal adversaries. These laws were the Export Control Act of 1949, and the Mutual Defense Assistance Control Act of 1951 (also known as the "Battle Act"). These statutes, enacted during the height of the U.S. policy of containment, reflected the prevailing view at the time; i.e., that any Soviet or Eastern European economic development would ultimately contribute to the military might of the Soviets and their allies, and was therefore to be inhibited, discouraged, or forbidden outright. Accordingly, we embargoed the sale of nearly all products and technology to the Soviet Union, its European allies, and the People's Republic of China.

But U.S. foreign policy, including foreign trade policy, underwent a significant change during the 1960s and 1970s. Business and other interests substantially increased pressures for the liberalization of trade with communist countries. This resulted in changing the substance of the Export Control Act. In its place, in 1969, Congress enacted the Export Administration Act. This statute, based on the belief that increased trade with the Soviet Union and its allies might ameliorate broader political conflicts, helped produce a substantial increase in East-West trade. The result was the hemorrhaging of U.S. technology into Soviet hands, which contributed heavily to the dramatic growth of Soviet military power. It also brought public recognition of the problem, as well as a shift in attitudes among U.S. legislators. Accordingly, in 1979 Congress enacted the Export Administration Act of 1979, which was a step in the right direction but is insufficient to redress the existing problem. The significant and positive new provision of the Act was a mandate to the DOD to develop an initial list of militarily critical technologies (MCTL), to be incorporated into the Commodity Control List (CCL), but at the discretion of administrative authority at the Department of Commerce. The MCTL has yet to be incorporated in part or as a whole into the Commerce Department's CCL.

An available alternative is to be found in the Arms Export Control Act of 1976 (AECA). This Act governs sales of technological data and hardware placed on the U.S. Munitions List. The Act relates to foreign military sales and licensed production or coproduction and recoupment of U.S. military research and development costs. The intent of the Act is to retard the transfer of weapons-manufacturing capability from the U.S. to other nations. The AECA may be suitable for safeguarding High Frontier core technologies in an international environment.

The participation of U.S. allies in the High Frontier's nonmilitary effort would add an international dimension. The nations of NATO have a tremendous economic and strategic incentive to cooperate. It is clear that our allies, in order to share the benefits, must accept the safeguards necessary to protect High Frontier core technologies and their space applications. In this international environment, we have a mechanism to achieve protection of core technologies: the 1976 AECA and the U.S. Munitions List. The AECA is most suitable and efficient because it provides adequate protection. At the same time, it is an established, familiar mechanism, accepted by the NATO alliance; and it provides for transfer, handling, and safeguards of military technology within the alliance. In other words, neither the procedures nor the relevant legal and security obligations represent anything new which might require adjustment.

Although the transfer of technology to communist countries has recently become a contentious issue both within the U.S. and among NATO partners, technologies on the U.S. Munitions List were not the subject of any dispute. Both classified and unclassified technical data and hardware of U.S. origin have been transferred to the NATO allies without significant problems. There are no complaints as to the terms and conditions which accompanied the transfers.

Arms Export Control Act and Munitions List

The provisions of AECA distinguish between classified and unclassified technical data and hardware. Nevertheless, any item placed on the Munitions List (ML) requires an export license which specifies policy conditions and technical level conditions of transfer. The constraints on the recipient are part of the license agreement; e.g., a German firm party to the agreement assumes the condition that the technology in question can be retransferred *only* to specified countries within the NATO alliance and *not* to those outside the alliance. Under provisions of the AECA, all items are, without exception, reviewed by the DOD. Furthermore, every item on the ML is automatically embargoed for export or reexport to a communist country. Lastly, one can track the movements of data and hardware.

Other provisions of the Act (not questioned by our allies) are that the U.S. has a free hand to write the terms of any transfer of technical data or hardware. Any transfer of classified items requires the recipient party to establish security procedures which are required of U.S. defense contractors handling classified data and hardware.

In the U.S., if a particular technology is placed on the ML, the owner of such technology is required to register with the Department of State, and to establish security procedures required by the DOD's Industrial Security Manual. Under the DOD security provisions for those who handle classified technical data and hardware, the subject must have his facilities certified as secure and participating personnel cleared for the handling of classified items. The Industrial Security Manual also stipulates that each firm handling classified material must have its own security officers who must work together with the Defense Industrial Security Clearance Office, which provides security clearance and maintains files on individuals who have such clearances in industry.

Much of High Frontier core technology would be classified. All security criteria which apply to domestic firms would also apply to foreign participants. The DOD, DIA, and CIA should prepare status reports on the adequacy and effectiveness of the security in a participating country or its industry. We reemphasize that our NATO allies have accepted the aforementioned criteria, conditions, and safeguard procedures. Consequently, we do not expect any significant problem in protecting High Frontier core technologies from Soviet penetration.

THE FRAME OF REFERENCE

For our purposes, the technology considered for protection must include arrays of technical information and know-how; keystone equipment and materials; and the accompanying goods. The criticality assessment of a relevant technology must be based on its strategic value. Upon determination of which technology must be controlled, this technology must be placed on the Munitions List, which is the only way to guarantee the fulfillment of requirements under national security considerations. By placing High Frontier core technologies on the Munitions List, we are also protecting against leaks of the same, via third party, because the U.S. will be the sole authority.

Special mention must be given to one particular

technology, microelectronics, which is the key to the High Frontier concept. U.S. and allied microelectronics technology is crucial to the successful implementation of the High Frontier. It is present in every aspect of core technology, and it is relevant to every system of the High Frontier concept — military and nonmilitary. Microelectronics is one of the main technological advantages the U.S. enjoys over the Soviet Union. It permits the reduction in size and weight of every relevant piece of space hardware and, at the same time, provides for optimization of capabilities and performance in terms of size and weight. It is estimated that the U.S. has a lead of some seven to ten years over the USSR in microtechnology.

CHAPTER VI
GROUNDBASED
ANTIMISSILE DEFENSES

In Chapter III, we emphasized the need for several layers of strategic defense. One of these layers— probably the first in terms of availability—is point defense, especially of the threatened landbased elements of the U.S. strategic deterrent. It is this particular threat to our Minuteman ICBM force that is opening the window of our vulnerability. It will be five or six years before spacebased defenses can firmly close that window. In the meantime we can partially close it by quickly deploying a point defense system that significantly reduces Soviet confidence in its ability to destroy a high percentage of our ICBM silos in a first strike. A quickly deployable point defense is a key supporting factor in the High Frontier concept. We have options available that could create, within as little as two years after the decision to do so, the minimum required silo defense—almost certainly at considerable cost savings over silo hardening.

During the past several years, many proposals have been made for rapidly deployable, simple, inexpensive groundbased ballistic missile defense (BMD) systems. Most of these proposals have been studied by various elements of the Department of Defense. In general, objections to these proposals have been that they are complex and expensive. In any event, funding of the BMD program has not been adequate to permit hardware experiments, so all of these proposed systems have been disposed of with paper studies.

Thus it could be said that our groundbased point defenses are, at present, paper empiricism. More needs to be done.

ESSENTIAL PROBLEMS OF POINT DEFENSE

For decades, a strong bureaucratic bias has worked against point defenses, i.e., systems which would defend a single U.S. ICBM silo. The best "cost-effectiveness" case was made for BMD systems that defend the largest possible number of assets with each defensive unit deployed. This search for cost effectiveness drives the BMD developer in the direction of *area* defense; that is, defense of a sizeable piece of geography containing a large number of potential targets.

But the more geography one attempts to defend, the more complex and expensive the antimissile system becomes, since the system must be able to engage numerous warheads simultaneously over a large area. The most difficult problem of all is that the farther away (and higher up) the incoming warheads are, the more difficult the problem of discriminating between real warheads and decoys. This area defense problem has not been solved to date—hence all decoys must be engaged as well. These factors both drive up the complexity and costs of area defense systems, and degrade system performance (kill probability) dramatically.

Worse yet, if an effective wide-area BMD system could be developed, the BMD system itself would become the most attractive and vulnerable target within the defended area. It is a lucrative target for the Soviets because its destruction is crucial to successful attack on the defended assets. It is also a tempting target for budget cutters; not only because of large total costs, but also because the entire complex system must be fielded before any military result can be produced.

The area BMD system is especially vulnerable to attack because of the softness of its critical radar components. They are subject to destruction by weapons of less accuracy and lower yields than those required to attack the defended assets. Given current Soviet submarine-launched missile characteristics, a Minuteman missile complex need not be defended against them. But if that complex included an area defense system, it would have to be protected against *all* Soviet ballistic missiles (e.g., submarine-launched missiles), raising once again the complexity and cost of the BMD system.

A point defense system, however, has some significant advantages over these area systems. Its required radars are relatively simple and inexpensive, and need cover only that small "threat cone" through which a warhead aimed at its single protected silo must come. A point defense system can operate almost autonomously; *automatically* when nonnuclear kill mechanisms are used. And a point defense becomes militarily effective on a unit-by-unit basis. Policymakers could choose to defend any number of hardened assets (e.g., missile silos) in any geographical distribution. Further, in the simpler point defenses, the problem of vulnerability of radar systems is solvable and the entire defensive system can be emplaced within the already secured real estate occupied by current U.S. ICBM silo installations.

TYPES OF SYSTEMS CONSIDERED

Both nuclear and nonnuclear quick-fix systems have been proposed, including such nuclear system ideas as planting nuclear charges of our own around our ICBM fields for detonation at an appropriate time to neutralize incoming reentry vehicles (RVs) with clouds of dust and debris. Other proposed nuclear systems include interceptor projectiles with small nuclear warheads. There has been a strong, persistent antipathy to nuclear systems by the DOD, Congress, and industry, who maintain that deployment would never receive popular support. Indeed, it would be difficult to muster political support for deployment of any new types of nuclear weapons on U.S. soil. Further, the use of nuclear warheads greatly complicates command and control problems. The nuclear defense system could not react without Presidential authorization.

The nonnuclear, quick-fix systems have generally incorporated interceptor projectiles of two types: those guided after launch, and those unguided after launch. Guided interceptors are considered more effective because, in the high winds and shock waves of a nuclear environment, their courses can be altered to obtain successive intercepts. The unguided interceptors could destroy one RV, but follow-on interceptors could not adhere to their predetermined courses in the highly turbulent atmosphere existing in the first few seconds following the first nuclear detonation. (These are generalized statements relating to carefully structured ICBM attacks.) The cost of an unguided interceptor system could be appreciably less than that of a guided interceptor system, and the time required for development and production could be less than that for the guided system.

The radars associated with nonnuclear quick-fix systems are generally postulated as relatively low power (10 to 20 kilowatt average power), with other design requirements well within current technology.

Typical required characteristics of quick-fix systems are that they can be sufficiently hardened against the blast and radiation generated by a one-megaton burst at about 5,000 feet altitude, and that they can be essentially unmanned and automatic in operation. The following systems offer the most promise of a quick fix. Details are available from proponent firms or government agencies.

Low-Altitude Defense System (LOADS)

This area defense system is funded in the defense budget and development has been underway for several years. Principal contractors are Raytheon (radar), TRW (software), Martin Orlando (interceptor), and McDonnell-Douglas (system integration). The system incorporates radars of modest power and guided interceptors with nuclear warheads. In later modifications it may incorporate a nonnuclear warhead, but only a token effort is underway on this feature. LOADS is included here because it has been funded, albeit modestly; early development efforts have already been accomplished. Test flights with hardware (except the nuclear warhead) may be possible somewhat sooner than the other systems described here.

Limited Area ABM System

This system is under study by Vought Corporation. It incorporates a phased array Patriot radar variant and guided nonnuclear interceptors with flechette warheads. The interceptor is a derivative of the Vought T-22 (Lance follow-on) missile having a 1,050-pound payload and Phoenix-type seeker. Designed to defend cities and high-value targets, the system could also defend Minuteman and MX.

SWARMJET

A concept under study by Tracor MBA, SWARMJET incorporates a radar system (using range-only radars), deployed in a trilateration scheme forward of the defended area. The radar information is fed to rapid-fire launchers with several hundred small projectiles per launcher. The projectiles are ballistic rockets of high velocity which can achieve a kinetic energy kill on an RV. The system has been under study for the defense of silo-based Minuteman. We will describe the SWARMJET system in considerable detail because those details are unclassified and may clarify operations of a point defense system.

The SWARMJET's radar emplacements are located some two to four miles forward of the Minuteman silo (see Figure 19) to detect, track, and calculate the optimum intercept point for the incoming RV. An acceptable radar system would include an array of three low-cost radar stations (UHF, VHF, or X-band) which determine the hostile RV's track and intercept point by trilateration. For trilateration deployment, the radars can be low-cost because only range information is needed. To prevent the necessity of hardening the radar against direct attack, a trilateration array should consist of four radar emplacements, and each Minuteman silo should be protected by two arrays. The radars in each array should be located far enough apart so that only one radar can be knocked out by a single nuclear warhead. Thus, it would take a minimum of two and perhaps four *perfectly aimed* enemy warheads to knock out the radar. The radar need not search above 40,000 feet altitude in order to provide sufficient tracking time (two seconds) to predict the intercept point accurately. Thus, the radars are not affected by high-altitude blackout, decoys, or clutter which may accompany high-altitude reentry by an RV. Because two arrays are located forward of the silo, there is no problem in looking around a nuclear fireball or even a low-level nuclear blast to detect other incoming RVs.

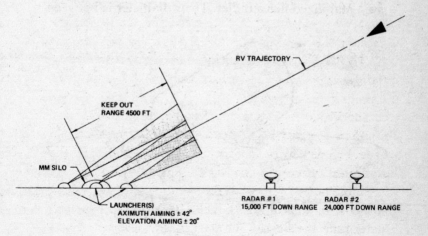

Figure 19. SWARMJET Engagement Schematic

The launcher system (see Figure 20) is hardened by either a concrete bunker with removable doors or individual steel shells, since the static overpressure from a one-megaton nuclear blast at 3,000 feet is a moderate 150 psi. The launcher slew rate required, 1.5 radians per second, is well within conventional launcher capability; and the aiming precision required, two to three mils, compares favorably with the one-mil precision often built into gun systems. Since each launcher would contain 500 to 1,000 small rockets, the preferred design is an open-tube (recoilless) launcher which minimizes the stress on the structure and on the launch tube weight. The launchers aim and launch rocket volleys in times of roughly one second. The rockets, about 10,000 in all, are literally a swarm of very small ballistic projectiles which fly to the intercept point at mile-per-second velocity to kill the RV.

The unguided rockets which make up the swarm are spin-stabilized like a bullet. Their diameter is between

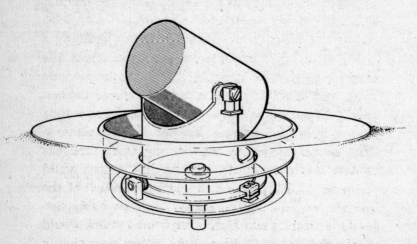

Figure 20. Hardened SWARMJET Launcher

one and three inches; their length, 10 to 15 inches. Using cases of conventional steels and one of the better ammonium perchlorate solid propellants, one of these small rocket projectiles can attain such high velocity that the kinetic energy of its impact against an incoming RV would exceed the energy needed to penetrate and destroy the warhead at ranges on the order of 4,000 feet—too far for an effective strike on our silo even if the enemy warhead detonates at that point.

Presuming an intercept point 4,500 feet from the silo, analysts calculated the number of rockets required to give an 85% probability of killing an RV—taking into account the uncertainties in the exact position of the RV and variations in the aim and flight pattern of the small projectiles. The necessary number of rockets to obtain this high kill probability is between 5,000 and 10,000. But these tiny rockets are exceedingly cheap because they are simple, small, straightforward, and can be manufactured in somewhat the same manner as small-arms ammunition.

One area of system concern is that of handling a highly structured, multiple RV attack in which the enemy can space the incoming RVs five to ten seconds apart, and equip RVs with salvage fusing or deliberately predetonate the first in the series. The winds generated by the first nuclear warhead generate a window of five to ten seconds, during which time it is unlikely that a successful SWARMJET launch could be made. An RV entering through the fireball of the first warhead during this period could probably successfully attack a silo. Such a structured attack would require detailed knowledge of the system operation; a commitment by the attacker of four or more weapons for each target attacked; and the perfection of highly sensitive salvage fusing for RV warheads.

Sandia System

This concept was developed by the Sandia National Laboratories at Albuquerque. It employs range-only radars in a trilateration scheme and unguided nonnuclear rockets which are directed to a point in space where the rocket warheads are detonated at the predicted RV location. The Sandia System application is to the defense of silo-based Minuteman or MX.

GAU-8 Gun System

Recent analysis by General Electric and Eglin AFB personnel suggests that at least one existing weapon could be deployed for point defense of a silo at very modest expense. This weapon is the high-fire-rate, 30-mm. GAU-8 cannon, currently installed as an anti-tank weapon in our A-10 aircraft. For point defense of a silo, the system would employ a simple, expendable range-only radar against RVs entering the threat cone at roughly 8,000 feet slant range from the gun. Assuming a single 500-kiloton Soviet warhead, our point defense must destroy or detonate it at least 1,000 feet from our silo to be an effective defense. The kill probability for a single GAU-8 system, firing at the RV between 8,000 feet and the minimum 1,000 feet, is very high—roughly 90%. This system is virtually off-the-shelf hardware and is so inexpensive that several hardened, pop-up installations could be deployed for lower cost than that of some other systems. The chances are quite good that a GAU-8 installation could, with its withering hail of small explosive projectiles, destroy several successive RVs. Even if the RV is salvage-fuzed—to detonate when struck—this defensive system is cheap enough to make that trade very attractive. With some modification to its target-acquisition subsystem, the GAU-8 could also defend a silo against cruise missiles; and certainly we have no shortage of inventive deployment options with a fully developed weapon that is already installed in aircraft. We also believe that the ABM Treaty, defining ABM systems in terms of launchers and missiles, does not pertain to this gun system.

SUMMARY

All of these systems achieve low-altitude intercept, except for the LOADS which proposes an intercept altitude of 50,000 to 75,000 feet Only LOADS has a nuclear warhead, and all systems are hardened against blast and radiation.

Although the need for ballistic missile defense is becoming more widely recognized, all but one of these quick-fix systems are today little more than paper empiricism. The problem now is not whether these paper systems would be effective, but whether hardware concepts (missiles, radars, launchers, C^3) are feasible in the areas of performance, producibility, and cost. To address this problem, we need greatly increased experimentation. The R&D budget allocated to such activities should be greatly increased over the next two years, the amounts being determined chiefly by the schedule on which prototype missiles and launchers can be produced. Radars should be as nearly off-the-shelf as possible.

CHAPTER VII
SPACEBASED GLOBAL
BALLISTIC MISSILE DEFENSE

A global ballistic missile defense (GBMD) system in space implies more than a network of defensive satellites; the complete system will require close support at orbital sites for inspection, maintenance, and patrol duties. These varied duties can best be performed by a small, manned, high-performance space plane (HPSP). The HPSP is detailed in this chapter as a vital element of our GBMD because it would permit us great operational flexibility. One dimension of that flexibility would be upgrading defensive weapons already deployed, as more advanced weapon systems become available for deployment in space. Therefore, we will conclude this chapter with a discussion of advanced weapons which we might eventually deploy in space via HPSP to supplement or supplant the current-technology weapons we describe here for GBMD-1.

The High Frontier's spacebased BMD system is designed to intercept groundbased, or sea and air launched, ballistic missiles such as the ICBM, IRBM, MRBM, and SLBM. The intercept is to be made with nonnuclear weapons, during boost and post-boost phases of the target missile's trajectory (Figure 21). We could also use this system for boost and post-boost intercept of other rocket-boosted vehicles, e.g.,

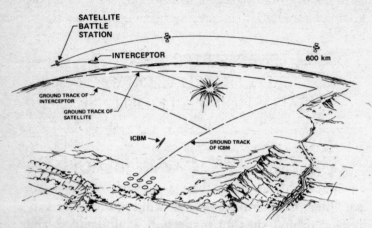

Figure 21. Boost Phase Intercept Geometry
This system uses multiple satellites armed with small
conventionally armed sub-missiles to destroy ICBMs and
other strategic missiles shortly after their launch. It uses
off-the-shelf technology and can be deployed in less than
five years.

communications satellites, intelligence satellites, or
assessment satellites. Our GMBD system includes a
large number of satellites launched in peacetime, and
each could be placed in orbit by a launch vehicle
consisting of the first three stages of the MX booster.
The network may carry tenant systems for uses other
than ballistic missile defense. Potential tenant systems include command, control, communication, and
intelligence (C^3I), nuclear detection, tactical warning,
and attack assessment systems. As an area defense
system, it inherently defends other forces as well as
ballistic missiles. Capable of defending cities and
countries, the GBMD system may provide defense
with international financial and political support.

Until now, ballistic missile programs and concepts

have been groundbased, and thus limited by the following characteristics and requirements:

- Defense of missile sites.

- Interception of each RV.

- Interception of RVs during the final phase of their trajectories.

- Need to discriminate between RVs, decoys, and other objects in free fall.

- Operation in severe nuclear environment.

- Dependence on optical and radar transmissions through the atmosphere in a hostile environment.

- Lack of realistic test capability.

- Inability to defend other forces, cities, or countries.

- Costs must be added to the ICBM and basing costs.

These limitations contrast sharply with the flexibility, performance, and cost effectiveness we can achieve with the spacebased GBMD system presented here.

Any spaceborne defense system must be able to survive in orbit. Performing their launches in peacetime, the ground launch facilities need not be hardened against attack. A single launch point may be sufficient for survivability as contrasted with the multiplicity of launch points for ICBMs or ground-launched BMD interceptors.

The GBMD system should be deployable in the near term, within roughly five years. It should have reasonable lifecycle costs — and lifecycle effectiveness, which correlates directly with survivability.

The GBMD system must intercept its target during the target's boost or post-boost phase. Its goal is to negate as many RVs as possible. Where RVs are MIRVed—fitted with multiple, independently targetable warheads—post boost-intercept can prevent subsequent deployment of warheads.

The High Frontier's orbital defense obtains its kills by nonnuclear impact at very high velocities. This kill method is similar to the intercepts planned with the USAF antisatellite miniature vehicle (MV) program, and the Army's homing interceptor (HIT) vehicle.

The system must not be dependent on the survival of groundbased C^3I or warning. This is not to deny that groundbased systems may assist the GBMD system. Rather, the GBMD system's independent capability is to assure its operation in autonomous modes for the sake of reliability. This requirement implies the uncertainties involved in our knowledge of C^3I during nuclear war.

To exploit the unique opportunity of the orbital system's location, its line-of-sight distances, and its remoteness, it may have other uses such as C^3I, tactical warning and attack assessment, verification, and nuclear-event detection. Such systems are termed "tenant systems". In other words, the GBMD system has tremendous—and cost-effective!—potential to carry tenant systems to enhance land, sea, air, and space forces.

The Soviet Union has deployed its ICBMs, IRBMs, and MRBMs across its country (see Figure 22). The missiles are also distributed rather uniformly from north to south in western Russia, while in eastern

Russia the missiles are deployed near the Trans-Siberian Railway. Our GBMD system will not be limited by the location, number, or distribution of the target missiles before launch. Similarly, mobile missiles and tactical or strategic reserve missiles cannot evade our GBMD system simply by their location uncertainty, numbers, or distribution. They must be launched to fulfill their purposes — and they will become vulnerable targets to our GBMD as early as boost phase of that launch, regardless of the launch point.

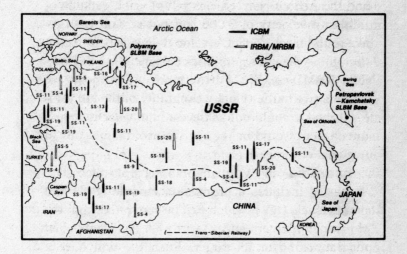

Figure 22. Soviet Ballistic Missile Distribution

In this discussion the Soviet SS-18 ICBM is referenced against our GBMD system. In Figure 22 these Soviet missiles are shown between Tyuratam and Novosibirsk. The SS-18 represents the greatest threat because it is accurate, each can deploy at least 10 independently targetable RVs, and it is being deployed in large numbers.

SYSTEM DESCRIPTION

A representative GBMD system includes a large network of satellites or "trucks", distributed in circular orbits at an altitude of approximately 300 nautical miles (nmi). The example here uses 432 trucks, all in orbits inclined 65 degrees with the equator. Target signatures of enemy missile boosters and buses are observed from our orbiting trucks at optical wavelengths appropriate for tracking with the Earth as a background. The truck can deploy 40 to 45 self-propelled devices called carrier vehicles (CVs), each capable of attaining a velocity of 3,000 feet per second with respect to the truck. This CV velocity is a critical factor, determining the kill capabilities designed into any particular GBMD concept. Furthermore, system costs are very sensitive to the velocity capability of the kill vehicle. (If the CV could not achieve a high velocity, many more of them would be required to avoid large gaps in our global coverage.) The truck tracks and supervises the control of each CV during its intercept trajectory.

Each CV includes a propulsion module (PM) and the kill vehicle (KV), which can be separated from the PM prior to intercept and after the KV has established optical tracking of its target. Each CV will have a pulsed light source whose pulse train identifies that CV uniquely. Thus, a number of CVs can be in the truck's field of view without causing ambiguity, providing a solution to the "traffic problem". The truck will be able to command each CV with signals for midcourse corrections, target designation, intercept inhibition, deorbit and burnup, etc.

Clearly, multiple strikes or intercepts can be made

from a single truck against one or a number of targets. Multiple deployment will increase the kill probability.

INTERCEPT GEOMETRY

If the CVs were deployed simultaneously in all directions from their truck, their distribution would be characterized as an expanding, approximately spherical surface moving with (and centered on) the truck, which is moving in free fall along its orbit.

Let us consider a representative cross-track intercept and make further observations. Figure 23 depicts the intercept of an SS-18 missile at the end of its boost from Tyuratam by a GBMD truck which is, at the moment, 300 miles above Saudi Arabia. Interception is predicted about 350 seconds from truck deployment, corresponding to CV deployment about 53 seconds prior to actual missile launch, when the truck is still roughly 950 nmi ground range from the missile launch point. If the truck were to move along its orbital path for 50 seconds, it could deploy CVs for final-stage intercept in response to direct viewing of the SS-18's launch. Ground-range separation between the target missile and the truck would be about 660 nmi for this example, and by this time the truck would be located over Iran. Trucks in orbits east of the orbit shown will also be able to intercept targets with CV deployments while over Afghanistan and China. Trucks in orbits west of the orbit shown may deploy CVs for intercept while the trucks are over the Mediterranean Sea, Turkey, or European countries. These interceptions would occur during the post-boost phase and could not prevent some Soviet RVs, deployed before intercept, from "leaking" through the GBMD system. Figure 24 shows that trucks can also intercept the Tyuratam missile near the end boost point, in essentially a head-on approach. These trucks are shown in Figure 24 as traveling down and to the right toward decreasing latitudes.

When a truck has moved along its orbit for 200 seconds' time, it will have advanced approximately 12.5 degrees along its orbit. The trace of the expanding envelope of the deployed CVs is indicated in Figure 23 by the dashed-line footprint (i.e., a two-dimensional representation of the effective kill envelope). The footprint shown is also valid at 50 nmi above or below the truck in its 300 nmi orbit.

SYSTEM DEPLOYMENT

Figure 24 introduces the complete global pattern of footprints. For clarity, only half of the footprints in the Northern Hemisphere are shown, and all footprints over the Southern Hemisphere have been omitted. Each of the teardrop shapes represents one footprint at an altitude of 250 or 350 nmi. Of course, the footprint at 300 nmi extends back to the truck. The apparent differences in footprint area are the consequence of the particular cartographic projection used (Miller cylindrical), and do not imply a real difference in footprint among the trucks.

The 65-degree orbit was selected for illustrative purposes because it allows us to present several points during the discussion which could not be illustrated if a polar orbit were used. Twenty-four orbits equally

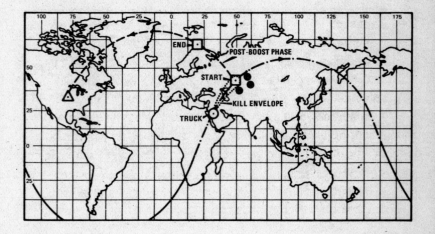

Figure 23. Cross-Track Intercept Example

spaced at 15-degree longitudinal increments along the equator constitute the full set. In this example (which is only one of numerous possible orbital geometries), each orbital path contains 18 GBMD trucks spaced in 20-degree increments. Thus, a total of 432 trucks are the full GBMD set.

Note that the 65-degree orbit causes substantial footprint overlap in the 45- to 65-degree latitudes. Small gaps or holes are seen among the footprints, which increase in area with decreasing latitude toward the equator. These holes pulsate in size as a function of time. The central pattern of the figure would extend uniformly across the hemisphere if all system footprints were shown. The complete pattern would be duplicated across the Southern Hemisphere as well.

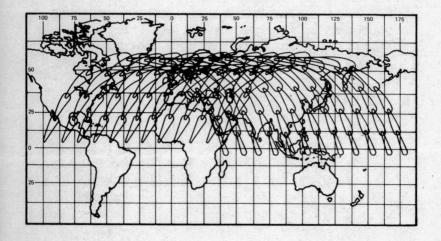

Figure 24. Example Footprints (1/2 Northern Hemisphere)

Figures 23 and 24 illustrate that the GBMD system has a second opportunity to intercept a missile strike against the U.S. by interception of RVs prior to entering the atmosphere. The RVs will be at risk while traversing truck footprints over North America, approximately 30 minutes after launch. The footprints over Canada and the northern U.S. continue the overlapping characteristics of those over Europe and the Soviet Union. The trucks in position to intercept these RVs will be (momentarily) located over the Pacific when the Soviet missiles are launched. We recognize that it is inherently more difficult to intercept the free-falling RVs without the plumes and heating effects of propulsion associated with rocket-booster and bus intercepts. In Chapter I, we pointed out that more advanced infrared sensors can upgrade our GBMD to the status of a GBMD-2, with the capability to destroy most free-falling RVs that "filter" through our first-tier kill envelopes. It is also proper to add that that the trucks attempting to intercept the RVs may be less vulnerable to attack than those which are, at the moment, operating over the Eastern Hemisphere.

LAUNCH

The truck is designed to be launched by the three-stage booster of the MX ICBM. The truck replaces the MX bus and acts as the necessary fourth stage to propel itself to orbital altitude and to provide the insertion velocity to circularize orbit. Subsequently, the truck's propulsion and reaction-control system will provide stationkeeping or orbit-adjust maneuvers, as well as evasive maneuvers if required to avoid attack. The truck contains a storable propellant and restartable liquid propulsion system with an axial rocket engine.

We expect that the trucks would be launched from a single facility on a Pacific island relatively close to the equator, such as Kwajalein, to take advantage of the Earth's rotation tangential velocity. This advantage also increases the payload to orbit over that obtainable if the same vehicle were launched into that orbit from the higher latitudes of the continental U.S. Rapid-fire sequential launches should not be difficult. Indeed, submarine-launched ballistic missiles can be fired at firing rates greatly in excess of GBMD requirements. The GBMD launch site need not be hardened because the trucks are launched in peacetime.

Tenant Systems

Once this system of interceptor satellites is in orbit—or for that matter, once any sizeable portion of them is in orbit—a sort of "geodesic dome" of kill envelopes is formed, essentially encompassing the globe. The satellite trucks will be constantly gathering information with their sensors and passing that information among them and to ground control stations. This creates an opportunity to make the GBMD system doubly valuable as a C^3I system. If the GBMD system is as survivable as the High Frontier studies conclude, this potential added capability can solve another of the gravest U.S. security concerns: the vulnerability of C^3I.

Survivability

The on-orbit elements of the system must be survivable through levels of nuclear war, and be capable of enduring with a residual performance capability. This is not to say that all elements of the system—or even perhaps the majority of the elements—must survive. The key point is that the *system* must be sufficiently survivable to provide an effective filter of ballistic missile attacks. The number of interceptors required for an effective GBMD will depend on such things as whether (for example) there is a second-tier terminal defense system, and the number of U.S. ICBMs which we determine must survive the attack.

We must consider passive as well as active defense options. Figure 25 presents, in matrix format, a row list of potential GBMD defense options or counter-

countermeasures (CCM) versus a column list of possible Soviet countermeasures (CM). To be realistic, we must include other considerations in Figure 25 determining whether a CM presents a significant threat to our GBMD. Among these considerations are:

- Level of Soviet technology required.

- Time to initial operating capability.

- Cost and other resource availability.

- Military impact and acceptance.

- Political impact and acceptance.

HIGH-PERFORMANCE SPACEPLANE

A truly military, piloted, high-performance spaceplane (HPSP) could perform a comprehensive set of space missions at all altitudes—from within the Earth's upper atmosphere to beyond the Moon. Its missions would include reconnaissance and surveillance; inspection and verification; antisatellite and anti-antisatellite; placement, supplementing, and standing in for unmanned satellites; on-orbit service, repair, and update of satellites; search and rescue; and missions requiring multiple atmospheric entry and exit. The HPSP transforms the Shuttle into an aircraft carrier in space, and extends its military operations throughout cislunar space (the volume of space between the Earth and the Moon). Compatible with ground and air launch, the HPSP would also operate completely independent from ground operations and from the Shuttle. The HPSP can be cached in orbit, brought back in a Shuttle, or piloted to a landing at unprepared sites, airfields,

Countermeasure	CCM I Shoot-Back	II Netting	III Survival Aids	IV Hardening*
A. Air-Launched ASAT	1. SOP 2. GBMD insensitive	1. SOP	1. Decoys 2. ECM	1. Potential — Weight limited to low (kinetic) energy attacks
B. Ground-Launched ASAT	1. SOP 2. GBMD insensitive to ASAT location	1. SOP	1. Decoys 2. ECM	1. Potential — Weight limited to low (kinetic) energy attacks
C. Ground-Based Laser	1. Not possible	1. SOP 2. Deploy CV's under attack	1. Decoys	1. Laser harden
D. Space-Based Laser	1. Not possible	1. Possible 2. Deploy CV's under attack	1. Decoys	1. Laser harden
E. Rifle-Shot	1. Not possible	1. Possible 2. Deploy CV's under attack	1. Decoys	1. Doubtful value
F. Pumped Belts	1. Not possible	1. Possible — Cooperative targeting of CV's	1. Not appropriate	1. Potential —
G. Spoofing	1. Not possible	1. Possible — cooperative source verification & validation	1. Not appropriate	1. Not appropriate
H. Jamming (Ground)	1. Not appropriate	1. Possible — cooperative	1. Not appropriate	1. Not appropriate

*Includes super hardened
box of CV's per proliferation

*Figure 25. Example Matrix of Counter-Countermeasures
Versus Countermeasures*

V Maneuver	VI Proliferation*	VII Replenishment	VIII Other Comments

1. SOP —
   Translations:
   Cross-range or
   along orbit (time)
	1. Feasible — Reduce
	   system sensitivity
	   to individual loss
		1. Feasible —
		   Use cached CV's
			1. SU can deploy A/C
			   away from SU/PACT
			2. Not evidenced
			3. Tracking quality req'd
			4. Small TW
			5. NUC CM fratricide

1. SOP —
   Translations:
   Cross-range or
   along orbit (time)
	1. Feasible — Reduce
	   system sensitivity
	   to individual loss
		1. Feasible —
		   Use cached CV 's
			1. Heartland launch
			2. Tracking quality req'd
			3. Large TW potential
			4. Limited LF's
			5. Drawdown ICBM LV's
			   & LF's
			6. NUC CM fratricide

1. Roll — versus CW laser	1. Doubtful value	1. Doubtful value	1. Pulse threat larger 2. Laser soft target 3. Locates laser (net) 4. Can proliferate
1. Roll — versus CW laser	1. Doubtful value	1. Doubtful value	1. Pulse threat larger 2. Laser very soft & few 3. Laser SOA?
1. SOP	1. Doubtful value	1. Doubtful value	1. Truck moving 24880 FPS 100/Sec $\approx \Delta$ S = 250 Ft 2. Rifle SOA?
1. Change altitude	1. Altitude, etc. differences		

1. Not appropriate	1. Possible — if
	   depleted use
	   cached CV 's
		1. Possible — if
		   depleted resupply
		   CV's only	

1. Not appropriate	1. Not appropriate	1. Not appropriate	Effective against RF sensors Low power laser could be used to jam EO/IR sensors wide beam to attack many simultaneously

*Includes replenishment
by cached CV's on-orbit
w/o truck complexity

aircraft carriers, or other ships. The HPSP differs considerably from other manned and unmanned space vehicles which have been proposed or studied. It differs in configuration, cost, performance, ease, and speed of development; and in the flexibility of its launch and recovery modes.

The HPSP would solve the problem of nonmilitary characteristics — and severely limited military capability — of current and proposed spacecraft, at a time when the military need for such a craft is substantial and increasing rapidly. Manned spacecraft programs and concepts are too often characterized by:

- Dependence throughout their missions on extensive ground-support monitoring, tracking, control, and communications.

- Extreme cost of acquiring, operating, and maintaining the ground-support and launch facilities and personnel.

- Vulnerability of launch facilities and global ground support to direct attack.

- Severely limited maneuverability in space.

- Narrowly limited mission profiles.

- Inflexible launch schedule.

- Weather dependency of launch and recovery.

- Little or no space rescue capability.

These limitations contrast sharply with the autonomy, flexibility, maneuverability, responsiveness, survivability, and cost effectiveness required of military operations, as established in military doctrine through long experience. Further, manned space vehicle programs have fostered the common belief that the economics, technology, and safety of man in space will force the continuation of these vulnerable characteristics into the future.

The National Command Authority and the DOD rely heavily on unmanned satellites as vital elements in C^{3}I, reconnaissance, and warning. Unmanned satellites have additional problems relative to manned vehicles. An unmanned satellite is inherently more vulnerable to attack; its utility is limited to a narrow range of missions; and it cannot think or adapt as a human can. Therefore, unmanned satellites compound the problem by limiting the reliability of their support to military space vehicles, and adding the problem of manned vehicles protecting, supplementing, or standing in for satellites. We must have balance and mutual support between our manned and unmanned military space systems. The manned vehicle must be capable of going "where the action is", including wherever the satellites may be in peacetime, conflict, and war.

In space, we need to provide the military man with a highly cost-effective vehicle system, with military characteristics that will help us secure the High Frontier. We would use these vehicles to:

- Protect U.S. resources from threats in (and from) space.

- Conduct aerospace offensive and defensive operations for the U.S. and our allies.

- Enhance the land, sea, and air forces.

- Serve as a practical utility vehicle in support of space assets and the exploitation of space.

- Support as many aspects of U.S. national policy as possible.

Studies to date have shown that the limitations of current designs are not inherent, and can be largely overcome by a new type of piloted military space vehicle. This high performance spaceplane would perform varied missions throughout cislunar space and in the upper atmosphere. It could do so with the flexibility and dependability we expect of an operational military system.

DESCRIPTION OF HPSP

An initial HPSP configuration is shown in Figure 26. To obtain high performance within the upper atmosphere, the vehicle is designed in the well-understood conical shape of the ballistic missile reentry vehicle. It is sized to carry one pilot, and its gross weight is roughly 6,000 pounds.

The extremely swept-wing conical vehicle is approximately 23 feet long and has a base diameter of 52 inches. A storable propellant propulsion system is standard in the basic vehicle, to provide propulsion for missions in low to medium altitude orbits. External tanks can be attached for increased velocity or range—somewhat analogous to the range extension gained by conventional aircraft when using external tanks. The engine comprises a circumferential ring of small thrusters (i.e., combustion chambers) around

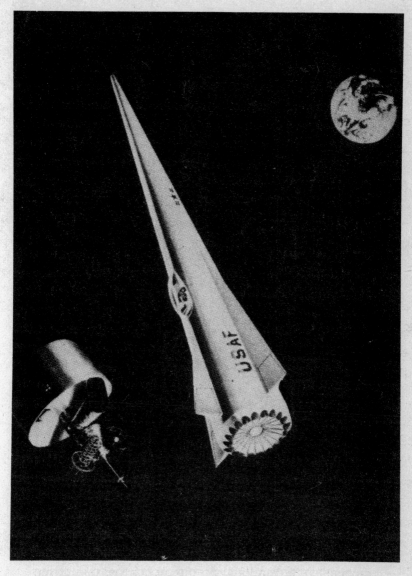

Figure 26. Military High-Performance Spaceplane
This vehicle has great flexibility in space and can perform a
broad range of military and nonmilitary missions.

the aft end of the vehicle. This nonconventional engine is called a "plug cluster" engine (PCE) and is capable of operating efficiently at all altitudes from sea level to the vacuum of space. Individual thruster control provides both throttling and thrust vector control for steering.

The entire vehicle is covered with ablative material over the surface of lightweight tiles. This insulation provides maximum thermal protection during reentry and permits distribution of the heat load by rolling the vehicle. We expect that the airframe will be made of a nonmetal composite material to take advantage of its low weight, compatibility with insulation tiles, and low cost. (It seems particularly appropriate that materials specialists expect still more advanced composites from R&D in space laboratories. Thus a first-generation HPSP would, in a very real sense, protect the development of its successors.)

During space flight and launch, the pilot can sit upright. During atmospheric maneuvers, the pilot is seated in a partially reclined position, and a hatch covers the retracted canopy. This restores the conical shape and provides protection from reentry heating.

The HPSP lands by controllable parachute. This recovery method permits piloted or automatically controlled flight to a low-speed landing at a small, unprepared site, an airfield, an aircraft carrier, or perhaps even a helicopter flight deck on a ship's fantail.

For very high velocity-change missions, such as rendezvous with a sequence of satellites or placement of payloads in geostationary orbit, an external propulsion module (or "extra stage") can be attached. Figure 27 depicts an HPSP in its extra-stage configuration. The external propulsion module is attached to the aft

Figure 27. Spaceplane with External Tanks

end of the spaceplane. The module contains an RL-10 engine and cryogenic (liquid oxygen and liquid hydrogen) propellant tanks. The RL-10 is the engine used most in space and is well suited to the HPSP. The circumferential overlapping of the spaceplane's aft end by its propellant tanks results in a shorter overall length than would be obtained with a conventional cylindrical propulsion stage with the spaceplane as its payload.

An alternative cryogenic stage, if length is not critical, would be a version of the Centaur upper stage. The Centaur also uses the reliable RL-10 engine.

The HPSP could perform an exceptionally wide variety of missions because it has the configuration and performance to exploit both space and atmospheric environments. It can be manned, yielding all the unique on-site capabilities of manned systems — but it also can operate in an unmanned mode.

Configuration Rationale

Let us briefly consider the logic that dictates this specific configuration of the HPSP. Derived from the fundamental principles of rocketry, orbital mechanics, aerothermodynamics, and hypersonic flight, this general configuration seems very likely to endure without much change in the foreseeable future.

Entry into Earth's atmosphere is necessary for autonomous operation as well as for optimum energy management and safety. Autonomous entry and recovery enables the HPSP to operate independent of recovery by the Shuttle. Proper energy management is vital to mission performance. Safety is vital to mission success — and, of course, to the pilot!

In terms of energy management, the ability to enter

and maneuver in the atmosphere gives the vehicle certain important capabilities.

1. It can extend mission range by maximizing the propulsive velocity available for mission tasks, in missions where less velocity is required to reach the atmosphere than to return to the Shuttle or some other rendezvous point.

2. It allows aerodynamic braking at perigee in the atmosphere, rather than depending entirely on rocket braking which requires more propellant. (Additional propellant would invoke a weight penalty and loss of subsequent maneuver velocity.)

3. Aerodynamic lift can be used to change the direction of flight (orbital plane change) with a return to space flight. This energy-efficient maneuver is called the "synergistic plane change," and is efficient for a vehicle with the lift-to-drag ratio and low total drag of the HPSP's conical configuration.

4. The HPSP can maneuver aerodynamically with negligible use of propellant to a landing point on Earth. This capability minimizes pre-entry propulsive maneuvers.

In terms of safety, the entry capability enables a recovery return, either to Earth or to a space station such as the Shuttle, depending on several variables. Such variables include the time available to reach sanctuary; the specific failure problem or damage that forced the premature recovery or abort; medical needs; and docking risks to the Shuttle. With sidecars, the HPSP can carry a number of passengers to and from the Shuttle, another manned vehicle, or a satellite. Without the atmospheric entry and landing capability, a manned orbital transfer vehicle—or any other manned vehicle—is not efficient, safe, or truly military.

The cone is the most understood and tested shape for reentry. It is the shape of ballistic missile reentry bodies for the same reasons as the HPSP, particularly the need for low drag and high lift-to-drag ratio. These factors result in minimum loss of velocity during atmospheric maneuvers. Therefore, the least amount of propellant is consumed in returning to space, and the maximum mission footprint or area is obtained in which the vehicle can operate. The cone presents the smallest surface area consistent with high aerodynamic performance; and surface area means weight in the thermally protected reentry body.

Minimizing vehicle weight is critical to propulsive maneuverability in space; to maximize vehicle payloads; and to the performance and size of the HPSP's launch vehicle (whether the launch vehicle is a Shuttle, an expendable rocket, or a launch aircraft).

There is no inconsistency between the argument for the HPSP's conical shape and the flattened delta shape of the Shuttle. The conical shape is correct for the generic, highly maneuverable spaceplane. Orbiter vehicles designed for maximum internal payload volume may require the winged, nonaxisymmetrical shape exemplified by the Shuttle—but they are penalized in performance. Because there is no drag in space, payloads for many HPSP missions can be carried externally; therefore the size and weight of the HPSP is minimized, resulting in optimal payload maneuverability performance.

Returning to the analogy of the ballistic missile, we observe that each missile pushes its payload (either the next stage, or the missile's ultimate payload) externally mounted for the best overall performance. We may conclude that the conical reentry body is best for vehicles designed for maximum payload maneuverability; high maneuverability with small internal payloads; synergistic plane changes; lightest weight; compatibility with launch by the Shuttle or the MX booster; near-term availability; and lowest cost. Other shapes may be best when internal payload volume is the overriding requirement. Vehicles of the Shuttle type may be characterized as logistic vehicles. That is, they are principally launch vehicles or payload-recovery vehicles for operation at lower orbits. Their use in higher orbits, or for high velocity change maneuvers, is not cost effective.

AIRCRAFT LAUNCH

Analysis of aircraft launch for the HPSP has shown that the combination of the Boeing 747-200F freighter aircraft, plus a rocket launch vehicle or booster, would place more than 20,000 pounds in low orbit. Therefore the HPSP, with a large payload and additional propellant in its external tanks, could be orbited using an existing aircraft as its first stage. Alternatively, two or three HPSPs could be orbited by one launch vehicle. Aircraft launch would be a flexible, military means for delivering large payloads to orbit where they could be transferred to various orbits by the HPSP.

The launch vehicle (carried aloft by the Boeing) is of conventional design using some off-the-shelf key components. A Titan LR-87 engine would power each of two strap-on boosters attached to a two-stage core

rocket, powered by a Titan LR-91 engine on the first
stage and three Pratt & Whitney RL-10 engines on the
second (final) core stage. Liquid oxygen and commercial liquid propane would be the propellants for the
Titan engines. Ease of handling, high performance,
and low cost are all obtained with this launch vehicle.

With the availability of either the MX booster or the
Shuttle as launch vehicles for the HPSP, and off-the-
shelf rocket engines for aircraft launch, the next step
is to develop the small HPSP to obtain its unique
cislunar maneuverability, its payload maneuverability,
and its omnimission flexibility. This vehicle can gain
and protect the "high ground" on the High Frontier at
comparatively minor cost.

Summarizing, we conclude that the HPSP would
bring great flexibility to our operations on the High
Frontier. The aerospace forces of the U.S. must be able
to go immediately to where its satellites are, and where
the threat is. The HPSP would, of course, have great
value in meeting nonmilitary emergencies in space. As
resources permit, the logistic-type Shuttle vehicle might
be further developed to operate under the protection of
the high-performance spaceplane.

ADVANCED TECHNOLOGY SPACEBORNE DEFENSE SYSTEMS

The High Frontier concept of layered defense visual-
izes two layers in space. The first layer to be deployed
will attack ballistic missiles in the early stages of their
trajectories; and can, to some degree, deny access to
space by other hostile objects. The second layer, using
more advanced technology, will be able to engage the

more difficult targets, e.g., individual RVs in mid-trajectory and hostile objects in near-Earth orbits.

These concepts of spaceborne defense tend to evoke images of Earth satellites armed with directed energy (beam) weapons, shooting at enemy missiles and at each other. Such images may become reality in the near future. But while beam weapons have been demonstrated in the laboratory, their deployment in a GBMD is too far in the future to meet the urgencies of the High Frontier. The High Frontier's second layer of spacebased defense, as we described earlier, will be based on improved infrared sensors which can improve our kill rates against RVs in the later stages of their trajectories.

Nevertheless, the ultimate promise of beam systems—especially in the light of intensive Soviet efforts to create such weapons—demand that a well-planned and funded U.S. R&D program be conducted, at least of coping proportions. Technological breakthroughs in beam weaponry might well tip the strategic balance at any time in the future. It is prudent, when faced with near-term defensive needs, to avoid *counting* on technological breakthroughs. We would also be prudent, when aware that a new weapon of tremendous potential is under development by an antagonist, to devote considerable study to the fledgling technology. Meanwhile, all requirements for the High Frontier layered defense can be met *without* the prior development of beam weapons or, for that matter, any other technological breakthrough.

Beam Weapons In General

Directed energy technology involves the generation of energy, and its precise and nearly instantaneous delivery to objects of interest at the long ranges required for space operations. Directed energy technology also provides opportunities for the U.S. to pursue valuable industrial and commercial operations in space.

Directed energy technology encompasses a family of similar concepts that collectively contain great potential for major military breakthroughs. These systems project intense electromagnetic energy (radio frequency, optical, X-ray) or subatomic/atomic particles (electrons, protons, ions) in narrow, precisely aimed paths. These energy beams can perform a variety of missions, including target tracking and destruction, electronic warfare, and surveillance. This family of systems includes:

- High-energy lasers (HEL)

- Particle beams (PB)

- High-power microwaves (HPM)

- Electromagnetic pulse (EMP)

Common characteristics of these technologies are the high propagation velocities of the energy, and the ability to focus that energy into a very narrow beam (hence the term "beam weapons"). Other characteristics of these beam weapons are:

- Effectiveness at extremely long ranges.

- Evasive maneuvers by the target are difficult.

- The weapon can engage several targets within a short time.

• The weapon usually has multishot capability.

Directed energy weapons operate on different principles than conventional weapon systems. Lasers, particle beams, and microwaves all project energy at or near the speed of light, i.e., 186,000 miles per second; but the beam ranges and the modes of interaction with both the targets and the intervening environment differ greatly.

Performance depends on a complex relationship between a number of system, target, and engagement parameters. For example: lasers might kill a given target by depositing large amounts of energy against its surface; microwaves might induce effects penetrating some distance into the target; and particle beams might penetrate deeper still. The kill mechanisms might be quite different in the three cases, and the total energy required for the kill might vary considerably, depending on the kill mechanism. On the other hand, the propagation losses between the weapon and the target might also vary greatly with the choice of the weapon, possibly making the easiest kill mechanism the hardest to achieve from a given distance.

For lasers and microwaves, the choices of wavelength and transmitter diameter determine most of the propagation characteristics. In addition, we must also consider atmospheric effects. Generally, a laser beam is more strongly affected by inclement weather (i.e., aerosols) than microwaves are. Atmospheric turbulence (like the shimmering of the air seen over a hot surface) also perturbs laser beams, requiring clever "adaptive optics" systems to correct for the disturbances. Moreover, there will be some combination of wavelength, beam diameter, and power level where the atmosphere will break down (like a spark), with great loss of power from either the laser or microwave beam. The ability to achieve energy levels high enough to kill a target varies greatly among the various choices. In addition to these factors, we must consider the nature of the hardware and its overall efficiency.

Particle beams using electrons within the atmosphere, or neutralized ions in space, are attractive. They have two potential advantages over electromagnetic beams, i.e., lasers and microwaves:

1. They are probably immune to atmospheric weather and dust clouds.

2. They penetrate deep into the target, making countermeasures difficult.

On the other hand, the required hardware may be very cumbersome; and the range of particle beams may be so short within the atmosphere that their use in strategic defenses is not feasible.

From the foregoing, it should be clear that the key to militarily useful beam weapons is the ability to deposit a lethal density of radiant flux against the target at maximum possible range. The levels of lethality may vary by many orders of magnitude, according to the nature of the kill mechanism. It is also important to note that all directed energy weapons have a long-range "soft" kill potential in addition to a shorter-range "hard" kill potential. That is, they can burn out or temporarily disable target electronics and/or electro-optical systems at ranges beyond the "hard" kill range.

The range and effectiveness of such kills vary widely, depending on the target and specific weapon and engagement scenario.

Directed energy technology has great potential for both military and civilian applications. Used in space, beam weapons may offer opportunities for worldwide projection of military force with essentially instantaneous destructive capability against satellites, aircraft, cruise missiles, ICBMs, submarine-launched missile boosters, RV buses, and surface targets. Landbased beam weapon systems—particularly high-energy lasers with spacebased mirror relay as well as high-power microwave devices—could also provide early capabilities against some of these targets. The impact of the development of these systems would be immense, definitely affecting the balance of world power. In addition, there are potentially important civilian applications, such as power transfer and propulsion. Here, directed energy technology could have important economic benefits in the longer term.

The potential mission which drives system requirements in beam weaponry the hardest is ballistic missile defense, against which 1,000 or more ICBMs may be launched simultaneously. This mission will stress most of the technology requirements. The BMD application has the highest payoff during the boost phase of ICBM trajectory, when the number of targets is minimum and the target vulnerability is maximum. If we destroy the carrying missile, we negate all of its multiple warheads. This phase lasts about 240 seconds —the burn time for the missile booster.

Another important potential application of beam weapons is destruction of high-flying aircraft. Aircraft at high altitude—above most of the diffusing layers of atmosphere—can be destroyed by spaceborne lasers of selected wavelengths.

Finally, practical potential applications for beam weapons include antisatellite use (ASAT), or defense of satellites (DSAT). The major advantage of beam weapons would be that they could very rapidly destroy hostile satellites or ASAT systems (for an illustrative example, see Figure 28).

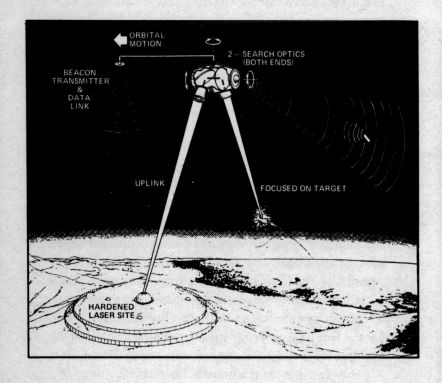

Figure 28. Advanced Global Ballistic Missile Defense
One example of a second-generation space-defense system
using a laser weapon.

Realizing the great potential of beam weapons depends in large measure on excellent command, control, and communications (C^3). Elements of command and control that must be considered in the development of beam weapons include operational control, tasking, monitoring weapon status, and assessment of battle damage. Also critical to effective beam weapon operation is the establishment of a coherent, nonambiguous space track picture of the surveillance volume. The global deployment of directed energy weapon systems will also require an extensive communications support capability that must survive and endure during all phases of conflict.

For the near term, the only directed energy weapons potentially useful for space applications are ground-based laser and microwave systems in an ASAT role. Either type can probably be deployed at a single ground site—preferably a site at high elevation to get above much of the diffusing effects of atmosphere. The technical risk appears to be low for the microwave system, but medium to high for the laser system, depending on the target range of interest. For example: the laser-system risk is medium for target ranges that include the present Soviet low-altitude ocean reconnaissance satellites; but high for longer ranges because sophisticated adaptive optics would be required. The microwave system suffers at present from a lack of development and agreement on the lethality of the kill mechanism. If the microwave power levels proposed are determined to be lethal, this system may be the better choice.

In determining the appropriate technical options for beam weapons, U.S. decisionmakers must evaluate the advantages and disadvantages of the various directed energy weapon technologies in a variety of deployment configurations. Particle beams, while highly attractive, are at a very early stage of development. High-power microwave technology has only recently started to receive attention as a potential beam weapon. However, our relative familiarity with microwaves and the potentially less demanding system requirements may result in its being used very early, if we favorably resolve the uncertainties about its lethality. Thus far, laser-system technology has received the most attention and is the most advanced. The several technologies which show promise are described below in some detail.

High-Energy Lasers (HELs)

Low and moderate energy lasers have made important contributions in numerous applications, including medicine, science, cartography, communications, range finding, and target designation. Potential applications for high-energy lasers involve not only weapons, but also laser radar, laser generation of fusion power, materials working (welding, heat treatment, etc.), and laser isotope separation.

The key components of a hypothetical laser weapon system include both the laser itself, which generates the high-power light, and the beam control subsystem, which aims the laser beam at the target and focuses it on a vulnerable spot on the target. Like other weapons, the laser weapon system must also have a fire control subsystem which acquires all the targets that need to be engaged; selects the one to engage; and tells the beam control subsystem where to look to find it. Finally, the fire-control subsystem decides when the target has been destroyed and designates the next target.

Shortly after the invention which made high-energy

laser beams possible, it became apparent that a laser damage weapon system, if it could be deployed, would have some particularly attractive features. For example, since light travels at 186,000 miles a second, the lethal flux would arrive at the target almost instantaneously. There would be no need to "lead" most targets. It takes six-millionths of a second for laser light to travel one mile; and in that time a supersonic aircraft traveling at Mach 2 will travel only a little more than one-eighth of an inch. Because of its pinpoint (rather than wide area) effects, a laser weapon could be used to selectively attack and destroy single enemy targets in the midst of a host of friendly vehicles.

A laser weapon can be expected to handle a large number of targets, even if those targets are coming from all directions. For each "shot" the laser takes, relatively small amounts of fuels are consumed to generate the beam. Thus, there is the potential for storing a large number of shots per installation (or a large magazine per weapon). Finally, since the beam is steered by mirrors, the laser weapon has the potential to move rapidly from target to target over a wide field of view.

Although such a system has enormous potential, development efforts must also address those characteristics of HEL weapons which tend to lessen their promise. For example: a successful laser engagement occurs only when the beam burns through the target surface and destroys some vital component (e.g., the guidance system), or ignites a fuel or warhead. Thus, while the energy is delivered instantaneously, the laser beam must dwell on the target to destroy it. Furthermore, "jitter" of the focused spot over the target smears the energy of the beam over a larger effective area, increasing the time required to damage the target. Thus the beam control subsystem must hold the beam steady on the target aimpoint. To do this, the target tracking and beam pointing functions of the beam control subsystem must be exceptionally accurate.

Fire control for laser weapons will have to be especially capable. It must be able to recognize and classify a host of potential targets, and to determine which one to engage first. In addition, to realize the firepower potential of a HEL, the fire control must be quick to recognize when the target being engaged has been damaged enough that it can no longer perform its mission; which means that the HEL can move to the next target.

A final example of a characteristic which tends to lessen laser capabilities is the effect our atmosphere has on a laser beam. As a function of the wavelength of the laser, the atmosphere absorbs some of the beam's energy; causes the beam to "bloom" or defocus; and adds jitter to the beam. Interactions between the high-power beam and the atmosphere effectively increase the spot size on the target, lowering the peak density of the beam and increasing the dwell time necessary for a kill.

Since in the vacuum of space the laser beam need not contend with degradations caused by the atmosphere, space has often been referred to as the "natural" environment for laser weapons. In this vacuum one can envision achieving the very long ranges needed if a weapon is to contend with the vast volume of near-Earth space. In addition, at long ranges, the stressing requirement of extremely accurate aim is reduced somewhat by the need for only low angular tracking rates. Thus, it may be easier to hold the beam on targets at the high velocities of objects in near space—typically 5 to 10 kilometers per second.

The high-energy laser scientist can envision a weapon in a high-density threat environment that methodically moves from target to target over its all-azimuth coverage; focuses the beam on the target; holds the selected aimpoint despite the target's speed and desperate maneuvering; burns through the target's skin; and destroys a vital component or ignites the fuel or warhead. Then, with instructions from its sophisticated fire-control system, the weapon switches its HEL to the next threatening target, and so continues through a number of successful engagements until its fuel is expended.

Although many different lasers were discovered in the 1960s, none was suitable for high-energy applications; some additional discoveries and inventions were needed. A principal discovery was that gaseous molecular lasers were possible. This discovery led to efficient lasers which generated their energy in the infrared portion of the spectrum. The next step was to invent a way to generate the energy required to operate the laser in an efficient and scalable manner. The required invention was made in 1967. It was the carbon dioxide gas dynamic laser—or CO_2 GDL for short. The CO_2 GDL was the first flowing gas phase laser that appeared to be scalable to very high energies, and it paved the way for serious consideration of a large HEL system. In recent years, other HEL concepts have been developed on this same basic principle, i.e., flowing gas, including the electric discharge and chemical lasers. Using these concepts, high energies have been generated at differing wavelengths.

The HEL program funded by our Department of

Defense involves development of many technologies, and is truly multidisciplinary. In addition to the usual scientific and engineering activities, special attention is devoted to the understanding of how a laser beam propagates through the air and interacts with its target. Moreover, in view of the possibility that our potential enemies may eventually develop a laser weapon system, we should devote significant resources to an investigation of techniques by which systems can be hardened to increase their survivability in a laser weapon environment.

The required operating time of an HEL is important, since it places demands on the power source as well as the aiming subsystem. If the laser is chemically fueled, then an adequate amount of fuel must be carried into space. Alternatively, if the laser is primarily electrically powered, then a potentially massive power-generating, storage, and conditioning system may be required—or possibly a small nuclear power system. Soviet emphasis on small nuclear power systems in space contrasts with an essentially nonexistent U.S. program.

The highest power chemical laser system currently demonstrated is potent enough to support early development of a landbased system for a limited ASAT capability. Somewhat higher brightness systems would increase the utility of groundbased ASAT systems. When configured for space operation, such a system would easily handle the ASAT requirements and could begin to handle more demanding engagements.

A spacebased chemical laser force could provide rapid global projection of U.S. power in limited conflicts. It could provide simultaneous continental U.S. and fleet air defense, and could be used to attack enemy airlift lines of supply and their airborne warning and control aircraft. This ability is considered unique, since no other system has potentially instantaneous global antiaircraft coverage. Spacebased chemical lasers of exceptional brightness are technologically feasible in that no insurmountable technical issues have been identified, and multiple potential solutions are possible for known critical issues. Deuterium fluoride chemical lasers have demonstrated the highest average power to date, although none of the chemical laser programs are specifically designed to demonstrate space laser technology. Our current efforts are aimed at developing new chemical laser configurations designed for scale-up to much higher power levels.

Another promising technological choice is the free electron laser which allows tunable wavelengths and promises high energy conversion efficiencies. The current free electron laser experiments are intended to verify the analytic models, and to demonstrate efficient operation using linear accelerators. These experiments will be continued to increase the efficiency to 30 percent. Further plans include wavelength scaling into the visible spectrum with higher energy linear accelerators, and linear accelerator stability experiments. We also plan to develop a moderate, average power, free electron laser to be completed in the mid 1980s.

Another possible technology which could lead to an effective beam weapon is X-ray laser. In space, much of the energy of a nuclear explosion is released in the form of X-rays, which are extremely short wavelength radiation. A 300-megaton blast in space could damage a satellite at a distance of several hundred miles. Thus, without any special techniques, a very large-yield burst in space might incapacitate a space vehicle. The USSR exploded a 60-megaton device in the 1960s and has boasted that the yield could be increased, but we have no evidence of such large weapons in the Soviet inventory today.

The abundant X-rays from an extremely high-altitude burst, i.e., in space, attenuate with distance. If a means can be found (perhaps through laser techniques) to focus even a small fraction of the X-ray energy of a nuclear device in space, the directed X-ray beam would constitute a formidable weapon against objects in space. A shielded object may be able to survive one X-ray "shot" but would probably succumb to a second shot.

Such an X-ray laser system would, of necessity, destroy itself to operate. It would have to destroy itself even to defend itself against a single kill vehicle. This creates an inherent self-contradiction for such weapon systems if deployed in space. Technologists reply that a single shot might be designed to generate more than one X-ray beam, so that at least tens of targets might be hit. But imposing simultaneous multiple-hit requirements, with attendant difficulties of fire control, would appear to impose such additional engineering difficulties that X-ray laser systems should not be counted on for any near-term defensive needs.

Particle Beams (PB)

A particle beam is a stream of highly energetic atomic or subatomic particles such as electrons, protons, hydrogen atoms, or ions. (By comparison, laser beams are composed of radiant energy photons.) An electron beam would resemble a lightning bolt. Presently, aside from potential applications as weapons, PB machines have potential use in inertial confinement fusion for energy generation, simulation of nuclear weapons, heating and welding, high-intensity microwave generation, geophysical investigations, energy transmission, medical treatment (e.g., cancer), and basic physics experiments.

There are three key components of a hypothetical PB weapon system. First, there is the source of the beam—the beam generator—consisting of a particle accelerator and its associated supply of electrical power, energy storage, and conditioning. The accelerators are similar to those used in research in elementary particle physics, except that currents in the beam weapon are much higher. Second, there is a beam control subsystem to aim the beam and to determine that the beam has hit its target. Last, the PB weapon must have a fire control subsystem which acquires all the targets that need to be engaged, selects the one to engage, and tells the beam control subsystem where to find it. Then the fire control system decides when the target has been destroyed and designates the next target. These fire control functions do not differ materially from those of fire control subsystems for more familiar weapons.

An appreciation for the damaging effect of highly energetic particles striking an object can be gained by noting the damage lightning can do when it strikes a tree or a house. (As a matter of fact, since the beam resembles a lightning bolt, technologists will often refer to a shot from a PB generator as a "bolt".) In high-energy physics, experimenters have long been aware of the ability of highly energetic particles produced by atom smashers to penetrate into materials. As the beam penetrates, it transfers some of its kinetic energy from the beam particles to the target material and, in addition, generates secondary radiation in the material, which can also disable the target electronics. If enough particles in the beam hit the target, the rapid transfer of energy to the material cannot be dissipated by it. Thus, the beam can cause a hole to be burned or melted into the material, or a fracture from thermal stresses as a result of the rapid deposition of energy. Another example of effects can be taken from discoveries in the early days of space flight. Energetic charged particles generated largely by the Sun are trapped in the Earth's magnetic field, thereby forming the "Van Allen" belts. These natural particle beams require spacecraft designers to build shielded and resistant satellites if flights in or through these belts are to be made without damage to such "soft" components as computers or electronics.

Thus, one can envision a weapon based on a stream of highly energetic particles that travel at nearly the speed of light. This stream of particles would penetrate the metal skin of the target, transferring a large fraction of the beam energy to the target. Initially, as the beam enters the target, it would damage electronic components and, as the beam continues to deliver energy to the target, ignite fuels and explosives and/or create holes in the target.

Particle beam technology is in the early research and exploratory development phases, and fundamental issues of feasibility still remain to be solved. The next major milestone in the program is the establishment of scientific feasibility by addressing the key physics and technology issues that, once resolved, will indicate whether PB weapons are practicable.

A PB weapon is a system which produces a high energy, small diameter beam of either neutral atoms or charged ions to disable the target. In weapons design, the choice of charged versus neutral particle beams depends on the deployment mode. Charged beams can only be used in the Earth's atmosphere at relatively short ranges. Neutral beams can only be used in space where the range-limiting effects of the atmosphere are not present and long-range engagements are possible.

For many reasons, neutral PB range requirements are generally on the order of 10,000 kilometers. If ranges of this magnitude are practical, then the system is very attractive, since a relatively small number of neutral beam platforms could meet a large number of space defense requirements.

High-Power Microwave (HPM)

The term "microwave", as we use it here, encompasses the frequency regime from just above conventional radar frequencies to just below most laser frequencies. This is a loose definition, and there is some overlap at the boundaries of the defined regions.

HPM weapon concepts are based on a number of emerging high-power source technologies which may lead us to substantial improvements in radars, communications, etc., as well as microwave radiation in weapons. Recent advances by both the Soviet Union and the United States indicate that it is now possible to achieve orders of magnitude increases in averaged and pulsed power output at millimeter and higher wavelengths.

In the past, the frequency domain between lasers and high-frequency radars (i.e., between micron and centimeter wavelengths) has not been available for widespread military exploitation. Although there were some millimeter and submillimeter wave sources, those sources were limited to very low powers. Newer technologies, generally based on relativistic electron beam approaches, can circumvent the power limitations and make available, for the first time, substantial powers in the millimeter and submillimeter regimes.

The key interest in directed energy technology with high-frequency HPM systems is the potential for either destruction of space vehicles or burnout of their electronic components at very long range. With the high powers now achievable, microwave impulses are potentially lethal, particularly against such targets as cruise missiles, remotely piloted vehicles, aircraft, and possibly RVs. Satellites and other targets in space can also be killed by rapid heating of structural and functional components, as well as by inducing currents which damage sensitive electronic components.

Like the Soviet Union, the U.S. is developing a high-frequency HPM capability. The Soviet results to date are very impressive, and it is generally agreed by most U.S. researchers that the Soviets enjoy a long lead in some technology areas.

The dimensions of previously proposed microwave weapon systems were very large. More recent considerations of HPM technology indicate the possibility of smaller systems which could be spacebased. The technical challenges for this type of system appear lower than for any other directed energy weapon. However, there is uncertainty and lack of agreement on the lethality of microwave energy at the power levels studied. Furthermore, the uncertainty is associated with details of the target design, since a major kill mechanism of the HPM is leakage energy getting into the target's electronics to cause the damage.

CHAPTER VIII
CIVIL DEFENSE

High Frontier emphasizes the need for active defense measures, both ground and spacebased, to achieve a layered defense. But in this context, Civil Defense (CD) becomes an important passive fourth defensive layer. For more than 20 years, the U.S. has focused on offensive measures, largely neglecting the defensive side of the strategic equation. However, with a new policy of Assured Survival, we must reevaluate national CD programs.

The very existence of spacebased antimissile defenses would vastly increase the value of CD, while drastically reducing its long-term costs. Our spaceborne systems would attack all hostile missiles regardless of where those missiles were targeted. It is quite feasible to reduce the number of warheads reaching our civil, as well as military, targets to under 10 percent. This filtering of the attack sharply alters the predictions of cataclysmic destruction. In the past, some of these predictions have been so dire that many people despaired of establishing any effective CD measures. With High Frontier defenses, we find that CD can become highly effective.

Our spacebased systems would defend against long-range ballistic missiles fired from land or submarines, but not against shorter-range submarine launches, or from bomber or cruise missile attacks. However, we could bring other defenses to bear against these latter threats. In any case, the severity of such attacks on civil targets would be greatly diminished. Furthermore, the bomber or cruise missile attack can be detected earlier, giving civil defenders hours of warning time—as compared to minutes.

As we indicate elsewhere, the costs of truly effective CD measures might exceed $1 billion per year in the absence of active defenses. Our critique of current CD programs, and our recommendations for improvements, are based on the problems of coping with attack when we have no active defenses on the High Frontier. If we reexamined our CD requirements assuming that our active defenses were deployed, we would surely find alterations in the scope and priorities of current CD programs. We have not attempted to make that reexamination here.

Under the policy of MAD, this nation's citizens remain unprotected hostages to the Soviet Union's steadily growing nuclear strike force. By contrast, the Soviets have taken substantial steps to preserve the lives of its population in a nuclear war. One might conclude that MAD may not be truly mutual when Soviet citizens have effective CD while Americans do not. Since the Soviets may reach this same conclusion, we find that a revitalized CD effort by this nation can in itself be a deterrent to nuclear war. Aside from the fact that protection of our citizens must be prime, we see that CD is a key element in the strategic balance because effective CD would reduce the possibility that the U.S. could be coerced in time of crisis. A spacebased BMD would go a long way toward defending our cities and critical installations; but until we obtain such a defense, we should give greater priority to conventional CD.

As early as the 1950s, the Soviets had an extensive CD program which included fallout shelters, planned evacuation of cities, storage of grain and other critical items, and perhaps most significant, required survival training for every citizen of the USSR. Indeed, we have copies of their survival training manuals—and they contain much that Americans need to know. In 1961, Soviet CD was transferred from civilian to military control under Marshal V. E. Chuikov. In 1972, CD was further elevated in overall Soviet strategic planning, and Colonel General A. T. Altunin was appointed commander of the Soviet CD program, with greatly expanded responsibilities. By their own admission, the Soviets regard CD as "a strategic factor" that will make "a major contribution toward victory". This was stated in a policy paper, *The Philosophical Heritage of V. I. Lenin and Problems of Contemporary War*, edited by General Major A. S. Milovidov and translated in a U.S. Government Printing Office document in 1974.

Thus, the upgrading of CD to strategic status for the U.S., as we recommend, has for many years been an integral part of Soviet strategic planning for its population and industry. This in itself should be reason enough for our government to scrap the discredited policy of MAD which has dictated, among other things, that American lives should *not* be protected through CD.

However, renunciation of MAD will not in itself reduce the number of American lives at risk, nor improve our capabilities to deter Soviet aggression. The dangers resulting from 15 years of unilateral U.S. adherence to MAD strategy (unilateral because the Soviets work hard to assure that their citizens are not unpro-

tected hostages) can be overcome only by protection of
our lives, property, and defensive forces. A viable CD
program, then, is a necessary part of decoupling from
MAD.

In the next few years, before active defense weapons
can be perfected and deployed, American lives and
essential recovery assets can best be protected by
rapidly creating a strategically significant, yet relatively low-cost, CD system. Effective CD should be our
first step toward Assured Survival. We can attain
Assured Survival if the U.S. deploys, very soon, both
an effective CD system and military defensive systems
capable of protecting America and its population.

ESSENTIAL REQUIREMENTS

The ultimate requirement of any strategically significant CD system is credibility: to the Soviet Union and
other potential aggressors, to the majority of Americans,
and to the allied and unaligned countries. The credibility of our preparations will depend in part on wide
recognition that CD efforts can save a great number of
lives both during an attack and in the following recovery period; and in part on the measures we take to
protect enough of our agricultural and industrial assets to yield reasonable hope that we can recover our
national power and our standard of living. CD preparations must also be widely recognized as making an
important contribution to the *prevention* of wars, to
obtain continuing support by the public. American
leaders must stress the positive, hopeful advantages
of serious survival preparations.

To avoid any misunderstanding here about the nature of global perceptions of our CD effort, we should add that the U.S. must truly obtain this level of CD—not merely claim it. Hollow claims, in an open society such as ours, would be perceived as hollow, and would be worse than nothing.

The time is very late for us to initiate strategically effective civil defense preparations. In future years, the difficulties are likely to increase. We should make a start soon, low-key yet resolute, and build rapidly toward a credible lifesaving, nation-preserving CD program. Once this war-surviving project is underway, it will signal that the U.S. has abandoned the demoralizing MAD policy and is determined to attain Assured Survival—just as the Soviets have already done.

CHAPTER IX
ECONOMICS: COSTS
AND OFFSETS

In preceding chapters, we presented the case for High Frontier efforts in space, as well as nonspace efforts essential to the change of strategy involved. This chapter addresses economic aspects of the program, based on an illustrative set of systems which are among our available options, and which would meet High Frontier goals. Cost estimates are in constant dollars. The estimated times required to reach initial operational status, as well as the cost estimates, are critically dependent on a special management system. This management system would minimize bureaucratic delay and is discussed in Chapter X.

GROUNDBASED DEFENSE SYSTEM COSTS

Any of several kill mechanisms can meet the restricted point defense requirement of the High Frontier. The partially tested SWARMJET system, discussed in Chapter VI, could meet the requirement to destroy Soviet confidence in a first strike against our silos. It is a simple system that fires a large number of small conventional projectiles, which form a barrier against a warhead approaching a U.S. missile silo at about a mile from the target. It could be described as "dynamic hardening" instead of as an antimissile system.

During manufacture, the conventional propellant casting process could be accelerated. By gathering available facilities into an ordnance-type production line, we could obtain production rates of five to ten million of these rocket projectiles per year. This is adequate to provide enough rockets to defend 500 silos with three swarms of 10,000 rockets each—a total of 15 million—within the required time. Based on using available launch drive systems, concrete bunkers for hardening, and 1,000 rockets per launcher, the total system costs are approximately $4 billion, including five years operation and maintenance, or $8 million per silo. This would defend each silo against the first three attacking warheads. For defense against attack from only two warheads, the cost per silo is only about $5 million.

The GAU-8 multibarrel cannon system might cost less; some of the other candidate point defense systems would cost more.

FIRST-GENERATION SPACEBORNE DEFENSE COSTS

The requirement for an initial spaceborne BMD system can be met by using off-the-shelf hardware to create a multiple-vehicle orbiting system, as described in Chapter VII. This system could provide protection to our allies as well as to the U.S. The multiple satellite deployment permits one satellite to defend itself and several others from attack. It also has the potential for forming the basis of a highly effective, secure system of command, control, and communications.

Since the system makes maximum use of existing components, it may be the least expensive and quickest available GBMD option. We could begin deployment of this system in as little as three years, and have it fully deployed in five or six years, at a minimum cost of some $10–15 billion.

The size and weight of the carrier satellite in this system is constrained by the limitations in throw weight of the MX booster, which is proposed as the method of insertion into orbit. Variations of this concept, using larger satellites with greater capabilities placed in orbit by Shuttle or larger boosters, may prove technically preferable. These variations might raise the costs to roughly $40 billion, and might delay the time to full operational status by about two years.

SECOND-GENERATION SPACEBORNE DEFENSE COSTS

The most promising possibility for a second-generation spaceborne defense is infrared sensor improvement of GBMD-1. This GBMD-2 system could be ready for deployment in 1990 at a cost of about a $5 billion add-on to GBMD-1 costs.

The requirement for higher-technology space defense systems might also be met by a high-energy laser system on the ground, with redirecting mirrors on satellites; by beam weapons deployed in space; or in hardened pop-up ground installations. These systems are currently being researched. Costs to continue research should probably be increased by about $100 million per year.

High-Performance Spaceplane Costs

There is an urgent need for a multipurpose, manned space vehicle which would perform a wide variety of missions. Some of these missions would be inspection of friendly or suspect space objects; satellite and space-station protection; adjustment or retrieval of satellites; and search and rescue. One such vehicle is the high-performance spaceplane, or one-man space cruiser, which would chiefly use available hardware and technology. It could be operational in several years for less than $500 million in development costs.

Civil Defense Costs

At present, the limited funds for a few Crisis Relocation Planning (CRP) area tests continue to be diverted to less disturbing, more publicly acceptable emergencies. The funding for fiscal year 1982 probably amounted to about fifty cents per person—and most of this annual pittance was spent on floods, windstorms, and other natural calamities instead of civil defense. In purchasing power, our current funding is about one-sixth of our 1962 CD budget. An effective CD program will certainly require healthier short-term funding, which would be subsequently reduced.

CRP should be greatly accelerated, its essential completion scheduled within three years of funding. A stretched-out, less effective CRP program than "Program D", called "D Prime", appeared to key CD supporters to be the best CRP program attainable. This stretched program was scheduled for completion seven years after an initial CD annual budget of $167 million was voted. To date, Congress has failed to appropriate even the modest funds needed to make a small start of such a minimal, stretched-out program.

The costs of mass-produced, austere concrete blast shelters could be as low as $200 million—a minuscule amount for the urban protection they would afford.

IMPROVED SPACE TRANSPORTATION

The immediate answer to improved space transportation is an upgrade of the current Shuttle program; we must improve turnaround time and create an unmanned, "cargo only" version. At the same time, we should begin development of a vehicle with much heavier lift capability, as discussed in Chapter VII. These improvements would cost an estimated $6 billion over a ten-year period.

MANNED LOW EARTH ORBIT SPACE STATION

A currently proposed military Space Operation Center should be given high priority. It should be expanded in concept to include provision for fly-along industrial/commercial space installations. The space station should be equipped to receive energy from a prototype SPS. A ten-year program to deploy this space station should cost about $12 billion.

SPS

We could place a solar power satellite in geosynchronous orbit using existing technology. With a microwave rectenna and conversion system on Earth, we could obtain 500 megawatts of continuous electrical power from a pilot system. This pilot version of an SPS, modified to power a space station by laser transmission, would cost about $13 billion.

SPACE INDUSTRIAL SYSTEMS RESEARCH AND DEVELOPMENT

The R&D costs for industrial space applications would probably be borne almost entirely by private enterprise, with little more than $50 million annually in government support.

SUMMARY OF COSTS

The options available to meet High Frontier objectives involve both space and nonspace systems, some conventional, some nuclear; and various types of beam weapons. System selection should be made by the task force recommended in Chapter X. Actual High Frontier costs, of course, cannot be determined until specific systems are selected. However, we have roughly estimated the costs of certain illustrative programs in the preceding pages.

A combination of military and civilian programs which cost about $20 billion over a five-year period, and about $35 billion through 1990, should meet the program requirements. We have carefully considered tradeoffs, in terms of current DOD requirements which High Frontier military systems would meet in part or in full. We conclude that projected defense budgets will increase only slightly *if at all*. Commercial and allied support of High Frontier nonmilitary systems could further reduce demands on the U.S. Federal budget.

Cost data validity for the illustrative High Frontier programs range from hard to very soft. In all cases, proponents of a program—private companies or concept authors—provided the data. This raises a possibility of bias on the low side of the cost estimates. Our analysis took this into account by rather arbitrarily increasing (in some cases as much as 100 percent!) the cost figures provided by basic sources. We did this despite the fact that we foresee major savings in estimated costs, if the acquisition measures proposed in Chapter X are used.

The costs estimated for several programs are much firmer for program totals than for year-to-year breakdowns. Further, the cost calculations take no account of inflation and appear in constant dollars (see Figure 29). Even with the detail uncertainties in our cost projection, it still demonstrates the feasibility of vigorously pursuing a set of programs at reasonable cost. The estimates in Figure 29 indicate that the first five years would cost $24 billion (about $18 billion for military and $6 billion for nonmilitary systems). Over eight years, the total costs reach about $40 billion ($27 billion military, $13 billion nonmilitary). Beyond eight years, roughly $10 billion more may be needed to operate and maintain deployed systems and to reach initial operational status for an advanced Shuttle, a pilot SPS, and an advanced space defense system. Thus, total commitment to this set of High Frontier programs could amount to some $50 billion over the next ten to twelve years.

High-frontier systems	IOC	FY1	FY2	FY3	FY4	FY5	5-yr total	(Military)	(Civilian)	FY6	FY7	FY8
Point defense of 200 silos	2 yr	150	1,040	30	30	30	1,280	(1,280)	(0)	30	30	30
Global BMD I (off the shelf)	5 yr	300	1,000	3,000	3,000	3,000	10,300	(10,300)	(0)	3,000	60	60
Global BMD II (advanced technology)	10 yr	50	50	100	300	1,000	1,500	(1,500)	(0)	1,000	1,000	1,000
High performance spaceplane (includes ACFT Launch)	4 yr	10	50	100	150	200	510	(510)	(0)	200	200	200
Shuttle improvement	10 yr	200	300	500	500	500	2,000	(1,000)	(1,000)	500	1,000	1,000
Space tug-orbital transfer vehicle	8 yr	16	56	234	325	325	1,156	(0)	(1,156)	253	205	200
Low-Earth orbit space station (military and civilian)	8 yr	37	138	1,364	2,946	2,337	6,822	(3,411)	(3,411)	2,023	800	200
Satellite power system (R&D plus pilot)	12 yr	30	60	100	150	200	540	(0)	(540)	200	1,300	1,800
Total by year (rounded)		800	2,700	5,400	7,400	7,600	24,000	(18,000)	(6,000)	7,200	4,800	4,300

Note: Millions of constant dollars.

Figure 29. High Frontier Estimated Cost Parameters

These figures could err seriously on the low side, but the military and economic consequences of success would make these programs a strategic bargain at twice the indicated costs. Alternate GBMD systems studied called for additional costs—but did not appear to double the estimates we cite here. The cited numbers are limited to acquisition and operation of systems. In overall strategic military cost, it is important to include the value of defensive systems as an "insurance policy" for expenditures already made and programmed for the next decade. The following analysis offers a different perspective which makes the High Frontier programs even more attractive.

THE ECONOMICS OF SURVIVABILITY

To put survivability in an economic context, we must consider how much the U.S. has spent on defense over the past decade (see Figure 30). To equip and maintain an Army with three-quarters of a million trained personnel has cost the U.S. a quarter of a trillion dollars (current dollars) over the last decade.

Our Navy with its ships, associated personnel, and equipment has cost a little more—about a third of a trillion dollars. These services, plus the Air Force at intermediate cost, have cost us nearly $900 billion altogether through the 1970s. This is the expense we have borne in order to be able to defend ourselves should the need arise. However, if the Soviets elect to commit to a nuclear strike, this enormous past investment on our part would be largely nullified. With the ability to destroy 99 percent of all bases they choose to hit, they could summarily destroy the majority of our facilities and our corps of trained personnel.

In simple terms: if we choose not to buy GBMD at $30 to $50 billion, the *hundreds* of billions we have spent on our broad-spectrum forces can be destroyed at will by Soviet military action, should they perceive that the stakes of the game justify that level of military risk.

As the current defense budget takes hold, the amount of DOD spending at risk will rise to still greater heights. For example, about $2.5 *trillion* in total DOD expenditures is contemplated through the 1980s.

Is it conceivable that the U.S. would forego an expenditure of only $30 to $50 billion to protect our $2,500 billion worth of defense purchases?

Figure 31 compares our options if we consider only the replacement cost of the facilities and equipment at risk in the event of a Soviet strike. The first column is a rough estimate of the value of all the targeted facilities (plus equipment such as planes or ships likely to be there), priced at replacement cost in 1981. These military facilities would cost about $725 billion if totally replaced. If we build MX–MPS, the total investment lost—shown in Column 2—to a successful

Soviet strike if we do not have ballistic missile defenses comes to $764 billion. This is more than the a priori value of $725 billion because the value of the MX investment is also included.

For all the defense cases following, only high-kill-capability systems are considered.

Point defense alone (100 percent coverage of Minuteman with a P_k of 0.8) reduces the Minuteman loss to about one-third of its total cost; but point defense suitable only for hard sites does not save any of the huge losses postulated for our soft targets.

With spacebased defense, our large DOD investments in bases and other targets can be protected — particularly if both boost and mid-course interdiction are implemented by a GBMD-2 follow-on. In this case, our facility and equipment losses decline to under five percent.

The final case, three layers of defense for Minuteman, cuts Minuteman losses to under a half-billion dollars. Note that it would not seem cost effective to spend $10 billion to trim only about $2 billion of Minuteman loss; and this would be true if point defense were added *after* the space defenses were in place. In actuality, point defense is essential as the first system implemented, because we can put it in place far more rapidly than the more effective GBMD systems. An initial point defense deployment is essential if we are to reduce our near-term vulnerability which is particularly acute.

To summarize our findings on the economics of military survivability, the Soviet Union has a nuclear strategic capability and an ongoing production capacity with no cutoff in sight. This permits them in this decade to threaten — or to use at will — to destroy the

CURRENT DOLLARS

	ARMY		Navy
16	Divisions	531	Ships
758,000	Trained Men	686,000	Trained Men
60	U.S. Bases	70	Bases
$255B	Cost (1970's)	$323B	Cost (1970's)
$645B	Cost (1980's)	$817B	Cost (1980's)

	Air Force		Total DoD
130	Squadrons		
559,000	Trained Men	$870B	Services
140	U.S. Bases	$124B	Other
$292B	Cost (1970's)	$994B	Cost (1970's)
$739B	Cost (1980's)	$2515B	Cost (1980's)

Figure 30. DOD Expenditures

majority of our expensive landbased retaliatory and conventional military forces. Our vulnerability will be especially acute by the mid 1980s.

Virtually all of the existing defenses purchased by our DOD through past expenditures, and virtually all the defenses to be acquired by the enhanced programs of the current Administration, will depend totally on "deterrence" working—if we choose *not* to implement ballistic missile defenses. But we can buy defense sufficient to greatly enhance Minuteman survivability for far less than by buying an equal capability through expansion of our missile forces. Point defense of Minuteman can be obtained quickly by expedited procedures. We can acquire effective defense for all military targets

MILITARY TARGETS

Case	Facilities Value (Replacement Cost) -1982$-	Cost of Facilities Lost				
		MX-MPS (200 MX)	100% Point Defense	Boost Phase Defense	Boost & Mid-Course Defense	-Tier Defense
	Ref	1	2	3	4	5
Minuteman	55	102	16	12	2	0
Air Bases	240	238	238	60	10	10
Army Bases	70	69	69	18	3	3
Naval Bases	150	148	148	38	7	7
Overseas Bases	40	40	40	10	2	2
Other (Nuc., Fuel, Etc.)	170	167	167	43	7	7
Totals	725	764	678	181	31	29
Defense Cost		50	10	25	30	40

Figure 31. U.S. Losses at Replacement Cost — Military Targets

(as well as U.S. cities and civil targets) through space-based systems at moderate cost. These GBMD systems are the only means known, short of keeping pace with Soviet missile proliferation, for denying the Soviets a unilateral first-strike capability.

The seriousness of our vulnerability dictates that the U.S. act. The High Frontier defense is critical to our survival. From a purely economic view, virtually all planned DOD expenditures are subject to the given threat; consequently, no expenditure is more important than one to counter that threat.

COST OFFSETS

We estimate very large offsetting savings from High Frontier programs. Clearly, costs are in the ballpark of the feasible. They offer more potential than some programs now getting surprisingly high long-range seed money.

The basic cost question should always be one of "opportunity cost," i.e., cost of the best alternative opportunity foregone. Ideally, economists select the lowest cost alternative for reaching any given objective. In practice it is seldom, if ever, feasible to describe alternatives of literally exact equivalence (in the military field, generally called "effectiveness"; in civilian endeavors, "benefits").

High Frontier military programs could begin with funds now earmarked for the survivability of strategic systems. Another source could be the reprogramming of funds earmarked for systems or support activities that would not be required, once the High Frontier's strategic defenses were in place.

Not yet knowing the High Frontier's *specific* systems or their initial operational dates, one cannot suggest reprogramming on a year-by-year basis. However, the Administration's decision on the B-1 bomber and the MX missile included a commitment to research and development of defensive systems to protect our ICBMs. Presumably it included necessary funds for this, and the implementation of High Frontier systems would satisfy this Administration goal.

With respect to reprogramming, we assert that the proposed GBMD would reduce, or eliminate entirely, the need to harden existing silos to accommodate MX missiles. Funds designated to superharden the silos could be diverted to spaceborne defense systems and to the point defense of ICBM silos as we described earlier.

If we pursue Assured Survival through a combination of spacebased defenses and point defense, we can solve the strategic force survivability problem. Without strategic defense, this country is left with only two options in the face of Soviet nuclear force. We can resign ourselves to the terrible absorption of a surprise first strike, or depend on our surveillance systems to let us launch on warning—or as enemy warheads arrive on target. We have already described the gloomy consequences of these two options, and the dramatic improvements when various defensive systems are added.

Funds now programmed to seek other solutions to this problem (such as MX deployments, rebasing our B-52 force, or pursuing alternative ICBM passive protection schemes) could logically be reallocated to the High Frontier's active defense programs. For nonmilitary or dual-purpose systems (e.g., a manned space station and a pilot SPS), development and acquisition funds should be provided largely from nondefense sources. Solar power satellites offer an extremely attractive prospect for large-scale, free-enterprise investment in a profitable space activity. While the costs of building and deploying an SPS network are large enough to require direct government participation, industry investments in SPS research would be encouraged if government guarantees were made available.

OFFSETS IN COMPETITIVE SYSTEMS

Other programs not clearly seen as alternatives to a given proposal, but subject to possible reduction or elimination, deserve attention. For example, an effective BMD might permit deployment of fewer and less hard ICBMs. Clearly, the costs of such indirect alternatives should be deducted, or included by subtraction, in the opportunity cost equation. Thus, the opportunity cost of some of the proposed High Frontier systems may turn out to be small or even negative.

Point defense costs might be offset, at least in part, by deferring MX silo hardening, pending future decisions on ultimate MX deployment modes.

The spaceborne GBMD system's first payoff would be as a filter against ICBM attacks in early trajectory. The combined filter effect of silo point defense and in-depth GBMD defense can be very high, as we have shown. Even if each system is "leaky", i.e., if the average single-shot kill probability (SSKP) is lower than we hoped, the two systems together would probably be highly effective. For example, even if each system has a 30 % leakage rate, *together they will save 91 % of the protected force*. If the leakage is as high as 50 % for each, together they will still save 75 % of the force—a huge saving in purely economic terms, as well as military and civilian terms.

A boost phase intercept system (like a preemptive counterforce attack) also degrades all other enemy missile attacks. To the extent that a Soviet attack might have included C^3 targets, it may be possible for the U.S. to save a significant part of the $18 billion recently proposed for C^3I improvements. In fact, the multiple satellites of the defense system may be able to enhance C^3 survivability by hosting a redundant space communications relay system which would be hard to interdict. It might also make possible a reduction of the programs under development for a wartime continuity of government, including the protection of

presidential successors and their support teams. To
the extent that the Soviets might attack military and
economic targets, there could be offsetting savings in
the hardening and dispersal requirements. There is
also the possibility that proposed (not yet implemented)
U.S. civil defense programs of $1 billion per year or
more might be reduced substantially over the long
term, if the active defense systems of High Frontier
are deployed.

Our allies could also be protected by the GBMD
system as a result of its potential effectiveness against
SS-20 intermediate-range attacks. This could simplify
the impending theater nuclear force negotiations. The
GBMD would strengthen—or rather, restore—the U.S.
strategic umbrella. Such a system might contribute
more to the reduction of nuclear proliferation incentives than all the fuel cycle controls, and nonprolifera-
tion treaty measures, taken together.

None of the High Frontier systems would protect
against a depressed trajectory attack by Soviet submarine-launched missiles against our strategic air bases.
However, the vulnerability of these bases to such attack becomes less critical if our ICBM force is no
longer so vulnerable to a first strike. This fact could
affect the costly proposal to move bomber bases in-
land as well as other reduction requirements designed
to reduce bomber-force vulnerability.

The savings provided by the high-performance spaceplane might include reduction in the number of standby
Shuttle vehicles, since the spaceplane could provide
repair in space. There may also be savings in the size
of various space vehicles because of the refueling potential of a spaceplane. Of special interest might be

simplifying crew changes, as well as maintenance and supply, for a low-orbit manned space station. Within the illustrative High Frontier programs, we might find a cost tradeoff between the functions of the spaceplane and those of the orbital transfer vehicle.

ECONOMICS OF SPACE INDUSTRY

What would happen to the economies of the U.S. and the Free World in the event of a major thrust into space by the U.S. beginning in 1983? Economic forecasting is not a rigorous, testable science like astronomy or physics; the answers to such a question are necessarily qualitative. We could make market surveys and return-on-investment calculations for specific space activities which we might conduct in the near future. But probably more convincing—and more useful—would be considerations of the overall economic structure of the U.S., historical analogies, and public psychology.

Space industrialization, including military activities in space, may be defined generically as the extension of human activities beyond the biosphere of Earth. These activities can be categorized in the hierarchy that follows.

PRIMARY ACTIVITIES

The extractive industries—hunting, fishing, agriculture, forestry, and mining—obtain the raw materials necessary for survival and for all other activities. Primary energy sources, such as fossil fuels, biomass, and uranium, are included here as well as hydroelectricity.

SECONDARY ACTIVITIES

The processing industries—manufacturing, construction, petroleum refining, ore smelting, food processing and packing, and generation of electricity from fossil and nuclear fuels—convert raw materials, or already processed materials, into more useful or versatile forms. Steel mills convert coal and iron into various grades and compositions of steel; rolling mills convert bulk steel into beams, plates, sheets, and wires; automobile factories convert sheets of steel into fenders; and so on.

TERTIARY ACTIVITIES

The service industries—transportation, communications, news media, basic education and job training, health care, insurance, banking, legal services, military defense, much of civil government, etc.—facilitate and support all other activities. Mail and telephone services support the railroads and airlines; banking supports telephone services; data-processing services make banking possible. Without these service industries, all extractive (primary) and manufacturing (secondary) activity would remain cottage industries.

QUARTERNARY ACTIVITIES

Activities carried out for their own sake, or for the personal satisfactions they provide, fill this category. These include socializing, sports, hobbies, much of higher education, cultural activities, pure research, fine arts including music, vacation travel, and so on. Generally, we work so that we can play. So on the whole, the purpose of all these primary, secondary, and tertiary activities is to make possible our pursuit of quarternary activities.

Historically, the earliest space systems were deployed purely as scientific research projects in the interest of national prestige; thus they were really quarternary activities, carried out for their own sake and exempt from normal "benefit-to-cost-ratio" analyses.

During the 1960s, however, we applied space technology to weather observation and communication satellites, so that space industrialization came to encompass some tertiary activity as well. The focus both in funding and in visibility, of course, remained on the Apollo program—a pure example of quarternary activity.

As satellite communication became more profitable in direct economic terms, launch services grew in importance. This profit motive justified development and deployment of such projects as the Shuttle, Europe's Ariane vehicle which is a direct economic competitor to NASA vehicles, and the Japanese N-family of vehicles; all this in a further expansion of tertiary (service industry) activity. The Shuttle is now scheduled for experiments and small-scale production in zero-gravity processing and for on-orbit construction and assembly—which begins our secondary activities in space.

Finally, we are now giving serious attention to energy production or harvesting in orbit; and to the mining of extraterrestrial materials, first for use in space and later on Earth. When the first such system comes on-line, we will be conducting the entire spectrum of activities—from primary through quarternary —in space as we do on Earth.

The key factor in this historical pattern has been the

cost of access to and from orbit. With the Space Shuttle, our transportation infrastructure has made a major leap forward, comparable in importance to the Erie Canal and the Transcontinental Railroad which fed the mushrooming of American progress during the 19th century. Our next generation of launch vehicles, presently on the drawing boards, can lower the cost-pcr-pound of access to orbit another 20 to 50 fold. This will permit us to develop yet another quarternary economic activity on the High Frontier: space tourism.

If space industrialization is merely the extension of economic activities into a new location, beginning just 100 kilometers above our heads, why bother giving it particular attention? Is there anything different or unique about it, justifying the apparently higher costs of doing business out there?

In economic terms, space *is* inherently different because of its unique attributes. It features weightlessness; abundant supply of cheap and nonpolluting energy; hard vacuum; easy access to a wide range of thermal conditions; excellent visibility of Earth; limitless room; isolation from the biosphere; and staggeringly vast mineral resources. These factors, taken singly or in various combinations, imply a contrast to Earthbound industry. Orbital facilities offer the potential for higher productivity; rapid (even exponential) growth; and the creation of new wealth from completely new resources.

Industrial Potential

Several studies in recent years have been performed on the economic potentials of various space industries.

Perhaps the most systematic of these studies, insofar as civilian prospects are addressed, are the parallel studies on space industrialization over the period 1980 to 2010. These were performed under contract to NASA's Marshall Space Flight Center by Science Applications, Inc. (SAI), and by Rockwell International. The SAI study developed a useful scheme for characterizing space industries, which were catalogued in the three groups of the following discussion.

Information

This includes remote sensing of weather; Earth resources; ocean surveys; communications; navigation and positioning; scientific space exploration; and data processing and storage in space.

Materials

The earliest activities in this area will be zero-gravity processing of materials brought up from Earth. Later possibilities include mining of extraterrestrial materials and smelting of asteroid ores for use in space.

Energy

Here, the main possibilities are harvesting solar energy in orbit for delivery to Earth via microwave or laser transmission, or by direct reflection of sunlight. Other concepts include relaying power between continents via satellite, and storage or disposal of hazardous wastes in deep space.

To these areas of civilian space industry, we must add the following three areas of military activity.

Force Delivery

This category includes antisatellite systems, laser battle stations, tactical projectile weapons, and GBMD systems.

Command, Control, Communications, and Intelligence

The C^3I area overlaps with the information category, including military communications and weather satellites. It adds surveillance and reconnaissance satellites, theater-wide navigation and communications from orbit, orbiting command posts, and launch-on-warning systems.

Military Support

A number of areas mentioned above can provide support to force delivery and C^3I operations, but some of these applications must be singled out for special consideration here. These include repair on orbit, upgrade on orbit, propellant and consumables supply facilities, and housing and medical facilities in space. Nonterrestrial mining of strategic materials may also be considered a support service.

The real productive capacity of our nation will increase to the extent that development and deployment of military space systems assist the development of civilian facilities, or stimulate technological innovations in productive systems; or to the extent that space-based military systems are less expensive than ground-based systems designed to serve the same goals.

HIGH FRONTIER ECONOMIC BENEFITS

Below, we address the economic effects of a High Frontier program from several different viewpoints using, in varying degrees, the framework presented above.

Direct Benefits

Hundreds of potential products and services could be provided by industries in orbit. These have been described in varying degrees of technical detail, but only a few dozen have been investigated for potential market size, revenues, and return on investment. The SAI study, previously mentioned, evaluated several promising items: information services, material-processing applications, and people-in-space activities; and projected revenues for these items over the period 1980–2010.

Table 1 lists the evaluated services and products. Previous studies of the market potential for energy from space (SPS only) were used, together with a new market analysis for night-illumination satellites. The SAI cumulative projection for specific products and services is shown below. Because this projection includes only some of the possibilities, it is likely to be very conservative. On the other hand, the particular products and services included may be poor choices that may not succeed for unforeseen reasons.

Several key points emerge from this market analysis. First, total revenues from industries in orbit can grow to significant levels in comparison with the entire present U.S. Gross National Product. These revenues

Table 1

Services and Products
Included in SAI Market Analysis

Information

Portable (wrist) telephone
National information
  services
Disaster communications
Global search and rescue
Air traffic control services
3-D holographic
  teleconferencing
Direct TV broadcast (U.S.)
Vehicle inspection
Nuclear fuel locator
Rail anticollision system
Personal navigation sets
Voting/polling sets
Urban/police wrist radio
Land and water resources
Electronic mail
Vehicle/package locator
Education by TV (U.S.)
Coastal anticollision
  system
Ocean resources
Power network monitoring

Materials Processing

Drugs and pharmaceuticals
Superconditioning
  materials
Fiber optics
Bearing materials
Semiconductor materials
High-strength magnets
Perishable cutting tools
Jewelry

People In Space

Space tourism
Entertainment
Sports
Space hotel
Movies

Energy

Solar power satellites
Night illumination

may increase to one or two percent of present GNP by
the turn of the century, and to four to six percent of
present GNP by 2010. The SAI study was carried out
in 1977–8, and already appears much too conservative.
More recent assessments of communication services
alone now suggest growth rates of 20 to 40 percent

through the 1980s for satellite communications, with 1990 revenues in the range of $50–100 billion annually — nearly five percent of present GNP.

Second, information services will remain, as they have since 1965, the largest profit center in space for the next few decades; although the mass in orbit and the number of workers in orbit will be modest.

Third, energy from space has a huge potential for growth, with growth rates in the next century estimated higher than for information services. (The market analysis includes *only* the U.S. market for SPS. India alone, by the turn of the century, could absorb the power output of 75 SPS units of five gigawatts each, if they could be built that fast.) If SPS is part of the total program, the total mass in orbit and workers in orbit would be immense.

Fourth, reflecting the structure of our economy today, the materials industries will be much smaller in total revenues than either information services or energy production.

Clearly, the potential is enormous. We are today in a position analogous to the investors in the Virginia Company early in the 17th century. The Virginia Company was chartered by the Crown of England as a profitmaking, joint-stock company. The company founders had a lengthy list of schemes that failed to make money in the New World; but the Virginia Company was still an economic success because of an unexpected development. Once the colonists established a toehold on the new frontier, they discovered that tobacco could be raised commercially for export to Europe. Doubtless some of the concepts considered today for commercial use of space will prove unprofitable. Nonetheless, we know far more today about the nature of outer space than the original shareholders of the Virginia Company knew about the New World. We too will encounter surprising and highly profitable uses of space surpassing our present imagination, if we only begin to establish a toehold on the High Frontier.

Economic Growth Through Innovation

The U.S. economy has been in difficulty for nearly a decade. A variety of causes has been postulated, ranging from changes in demographic structure to unfair foreign competition and overregulation of our industry. While most postulated reasons have some validity, and some have been addressed by attempted remedies in previous administrations, little more than lip service has been given to the loss of our industrial productivity due to declining innovation in the technological base of our industries.

The economic growth of the U.S. during the period from about World War I until the OPEC embargo has depended almost entirely on technological innovations made in the latter part of the 19th century. Industries which today form the bulk of our economy, and are rooted in that age, include steel, electric utilities, automobiles, aviation, radio and television, petroleum, and chemicals. During the economic growth of this century, progress has been essentially continuous, relying on incremental improvements in these basic technologies rather than on fundamental, qualitative breakthroughs.

Peter F. Drucker describes our present situation as an "age of discontinuity" (*The Age of Discontinuity: Guidelines to Our Changing Society*, Harper & Row, 1968, 1969):

Genuinely new technologies are upon us. They are almost certain to create new major industries and brand-new major businesses. The growth industries of the (past) half century derived from the scientific discoveries of the middle and late 19th century. The growth industries of the last decades of the 20th century are likely to emerge from the knowledge discoveries of the first 50 or 60 years of this century: quantum physics, the understanding of atomic and molecular structure, biochemistry, psychology, symbolic logic. *The coming decades in technology are more likely to resemble the closing decades of the last century, in which a major industry based on new technology surfaced every few years, than they will resemble the technological and industrial continuity of the past 50 years.* (Emphasis added.)

By deliberately targeting investments into key areas of technology to take advantage of new opportunities in the marketplace—including those areas of industry in which real productivity is halted or declining—we can initiate a cycle of major economic growth. Of the industrialized Free World nations, only Sweden has done this systematically, which resulted in the recent transformation of Sweden from a primarily agrarian society at the end of World War II into a highly industrialized nation. Their major export industries now include automobiles, where Volvo and Saab pioneered the combination of consumer styling, low fuel consumption, and mechanical durability; and aircraft, where Saab and Viggen pioneered short-field fighter aircraft.

The key question in the context of "reindustrializing America", then, is this:

What new technologies offer the greatest potential for significant economic growth in which existing knowledge and techniques give the U.S. a commanding lead over potential competition from abroad?

Several answers spring to mind immediately: computers, genetic engineering, and space technologies. The first two have already taken off, largely in the private sector, since the front-end capital investments are comparatively quite modest. Pocket calculators now cost less than slide rules used to cost and are far more versatile, increasing the productivity of those who once used slide rules. (Slide rules have not been manufactured in the Western industrialized countries for about eight years; they have been displaced by pocket calculators.) In addition, a whole new industry, video games, has capitalized on this technology, with total revenues of more than $1 billion annually. The potential impacts of new computer technology are only beginning to be felt throughout the economy. Genetic engineering is still at a very early stage, but its impact on such industries as food production and on costs of pharmaceutical products and industrial chemicals will be enormous.

Space technologies, on the other hand, have not yet reached the takeoff point. This is due in large part to the high costs of transportation to orbit. Other factors include investor uncertainties about markets; perception of extended delay before an investment return; perceived and real regulatory inhibitions; and industry's perception that U.S. foreign policy has not reached a posture of clear support for private enterprise in space in the face of potential opposition and hostility from abroad. Most of these disincentives can be overcome, at no direct government expense, with a simple decision: *make a national commitment to the High Frontier.*

A coordinated High Frontier program in which civilian and military space systems are encouraged to share would provide the focus for targeted development of this major new technology. Technological development should be guided by specific marketable products or services to maximize economic benefits. Development of large space structures, for example, should lead directly to applications such as direct TV broadcast satellites and solar power satellites. Development of life-support systems for space should be planned for materials-processing facilities, on-orbit repair facilities, and survivable command posts in orbit. As in the case of our microelectronics revolution, these technological innovations would find countless unforeseen applications elsewhere in the economy, with positive effects on our productivity and thus on economic growth.

Would the investments needed for this program be better spent in some other area? Certainly this is a debatable point, but several considerations strongly favor space technologies. In the short run, most of the investments will be channeled into the existing aerospace industry. Of all U.S. industries, only retail sales employ more people per dollar of capital. Expanding the size of our aerospace industry would create more direct jobs than virtually any other allocation of effort. Previous studies (Chase Econometrics, for example) have shown very high "multipliers" for aerospace investments as well. Each dollar spent on the space program during the 1960s, for example, generated $6 or $7 of new GNP over the ensuing few years, supporting other jobs in the economy as well.

In the long run, no other technology in our grasp offers access to such huge new resources or to the prospect of creating totally new wealth from sources outside the present economic system. An investment in space during the next 5 to 20 years can provide access to the resources of the entire solar system — not just for the U.S., but for all the world. If the U.S. does not take the lead, those resources may well be developed by others, notably the USSR, with far less willingness to share access.

We should not allow energy price competition in the U.S. to blind us to the attractiveness of exporting energy from solar power satellites to the underdeveloped world. Since the entire economic infrastructure is very inadequate in the less developed countries — especially including the transportation industries — costs of energy there are typically three to five times higher than in the U.S. The rise in OPEC petroleum prices has aggravated the energy crisis faced by these nations. Building nuclear power reactors in Third World countries is much more expensive than building the same reactor in the U.S. because many of the parts, which are inherently high-technology products, must be imported via inadequate transportation systems. Most of the construction workers must be highly skilled and thus must also be imported. Any savings on energy costs, which might otherwise be expected on the basis of the difference between nuclear-generated electricity in the U.S. and alternative energy systems in the Third World, are severely curtailed by higher installation costs, as well as political difficulties.

In the case of the SPS, however, more than half the cost of each generating plant is attributable to the space segment. No cost differential applies between a highly industrialized country and a developing country. The groundbased receiver antenna (rectenna) system consists of very few high-technology components, e.g., the solid-state rectifier elements. These could be manufactured with highly automated equipment in urban factories in developing countries, with limited numbers of skilled workers. The overwhelming bulk of components for the rectennas could be assembled by local semiskilled workers. Construction of the rectenna could be done on-site by large numbers of unskilled workers with semiskilled supervision. Thus the capital costs for a complete SPS would be only slightly higher in a developing country than in the U.S. Third World energy costs would then be substantially lower than present sources such as petroleum, firewood, or cow dung.

Military Considerations

A national commitment to the High Frontier program would harness technology developments aimed at military purposes to support nonmilitary economic productivity. Economic benefits are certain, but difficult to estimate in quantitative terms. More importantly, if the strategic purposes of the U.S. can be achieved as well (or better) by going into space rather than by relying on more conventional groundbased systems, then the Federal budget can be reduced by the difference in system costs. This in turn would reduce government borrowing during the next few years, easing pressure on interest rates. The effects on the overall U.S. economy of lowered interest rates are discussed below. More specifically, however: if reduced Federal demand for borrowing eases interest rates, the cost to

the Federal government of servicing the national debt
(now roughly $1 trillion) would be reduced by $10
billion annually for each percentage point decline in
interest rates.

Economic Effects of Renewed Hope

With all the attention devoted in recent years to
perceived scarcities of clean air, clear water, petroleum,
natural gas, strategic materials, investment capital,
and government funds, we have barely noticed that
the most critically scarce commodity of all has been—
hope. In the growing expectation that things are only
going to get worse, consumers have had little, if any,
incentive to save. Consumer debt has been climbing at
ever faster rates. Unemployment among teenagers has
convinced much of American youth that the future is
bleak, contributing to a major crime wave and to
increased drug usage, with very large economic costs
to all society. The one segment of the public which is
resolutely swimming upstream, struggling to get completely out of debt, is the survivalist movement: people
who are becoming convinced that a major economic
and possibly social collapse is imminent.

A commitment by the U.S. to strong use of the High
Frontier could dramatically change public attitudes
and instill a new sense of purpose and hope in this
nation—as well as the rest of the Free World. A major
thrust into space would provide the world with clear
and convincing evidence that the resources available
to the human race are not fixed; that new wealth can
be created without depriving others. With renewed
confidence, consumers would be motivated to save for,
and invest in, the future. The prospect of new jobs in

space industries, even of jobs for skilled hardhat workers in orbit, can be expected to improve the morale of unemployed youth and to renew their faded dreams.

Public Psychology and the Economy

In this discussion, much emphasis has been placed on the economic effects of changes in public psychology. The role of these psychological factors in economics can be highlighted by the following anecdote. Several years ago, Eric Burgess, one of the founders of the British Interplanetary Society, was discussing some ambitious future space project with a very wealthy financier. When the financier asked his inevitable question—how much would the project cost?—Burgess apologetically quoted a huge number and added, "Unfortunately, it's just too expensive to afford."

"Nonsense," the financier retorted; "we *invented* money, didn't we?"

A national commitment to the High Frontier, if carefully presented to the American public and to the Free World community, could quickly alter the public psychology, especially in its view of the future. Such changes would alter economic realities for the better, more rapidly than any amount of tinkering with social programs or Federal Reserve discount rates. Some of these effects are discussed below.

Prompt Economic Effects

Two important short-term economic effects can be expected to result from a major national commitment to the High Frontier: increased savings and investments; and a softening of world energy prices, especially for petroleum. These effects would arise from changes in public psychology rather than from shifts in cash flows in the economy, or from actual growth in productivity. Since changes in public psychology can take effect within as little as a year, we should make major efforts to display a high degree of national commitment to the program, to guarantee that these effects materialize.

The going interest rate is determined by the anticipated inflation rate, and by the average discount rate. If either rate can be lowered, then the interest rate will inevitably decrease as well. When a potential lender or investor has disposable funds, he can either spend them immediately on personal gratifications (e.g., trips, new clothes, restaurants, luxury gifts, a newer or more luxurious car or home, etc.), or he can invest the money, deferring that spending until a later date. An inducement must be offered to the lender in exchange for the deferral of his gratifications. That inducement is the discount rate—and each potential lender has his own discount rate. When the demand for loans rises, a larger number of those lenders with high personal discount rates become willing to lend their money, rather than spend it on immediate gratifications. The average discount rate, then, is the rate at which the supply of investment money, and the demand for that investment money, balance to clear the market.

The expectation of technological innovations resulting in superior products or services at the same cost, or the expectation of increased productivity, lowers the discount rate for all investors. The result is that more money is saved or invested. More investment capital becomes available for reindustrialization, while interest rates stabilize or even decline. The availability of more capital for plant expansion, or for replacement of less productive equipment, would improve productivity throughout the economy. This further stimulates higher expectations, and further lowers personal discount rates for potential investors. An expectation of technological advances, and the resulting incentive to invest capital, would almost certainly result from a vigorous commitment to the High Frontier.

If the High Frontier program included a strong commitment to power from space on an economically significant scale, OPEC petroleum prices would soften significantly—and soon. Petroleum reserves in the OPEC nations are large, but finite. Prices and production levels are set to maximize the total accumulation of money from petroleum sales during the few decades before the cheap supplies run out. If it became clear to oil-producing countries that the demand for oil would decline significantly in 10 years or so, the value of oil retained in the ground would decrease; production levels would increase; and prices would stabilize or decrease.

This effect, even more than the improved investment market, would depend on a convincing U.S. commitment to a major new energy source which promises to be economically competitive.

The benefits of a softening or decline in world energy prices are fairly obvious. Less money spent for each unit of energy means that more energy can be obtained and used for the same outlay of funds; or that money is freed for other, more constructive purposes. In the industrialized nations of the Free World, energy consumption would probably rise a bit in response to the price changes, with most of the change in funds becoming available for other uses. In the less developed countries, consumption of oil would accelerate; they would now be able to afford the oil. This would strengthen the industrial base in the developing countries, with major benefits to both their agriculture and their environment. These benefits would accrue because, during the last decade, scarcity of affordable fossil fuels in these regions has resulted in deforestation and use of dried animal dung for fuel. Thus there is little fertilizer for crops, and few forest products for industry. Major banks in the U.S. have lent massive sums to developing countries to finance their petroleum imports. Billions of dollars of these loans are in risk of default—even with low interest rates subsidized by developing agencies such as the World Bank. The softening or decline of petroleum prices, by the strengthening economies of developing countries, would lessen fears in the financial community of defaults on these international loans.

Impacts on Jobs and Taxes

The SAI study forecast some jobs created and new tax revenues generated for the Federal government under three different assumptions about the space industrialization programs over the next three decades. The three scenarios were: the baseline case, with no SPS; the baseline case with SPS; and an "Upside" program, in which foreign competition and some incentives to private enterprise result in a vigorous civilian program.

Both baseline cases, just four years later, already appear hopelessly conservative due to explosive growth in commitments for communications satellites. The High Frontier program would encompass a higher level of national commitment (due to inclusion of the military systems) than was envisioned in the "Upside" program. (The SAI study was conducted under ground rules which specifically forbade considerations of military activities in space, let alone synergisms between civilian and military space efforts.) Given all these caveats, Table 2 shows these conservative projections.

The SAI study comments on these results as follows:

The estimate of jobs for 1985 is probably low by a factor of two since most funding would be to the aerospace industries. The Aerospace Industries Association (AIA) has estimated that about 30 direct jobs are created for each one million dollars of appropriation. Direct plus indirect jobs are estimated to total about 100 jobs per one million dollars. Thus, the job projection for 1985 is conservative. . . . The true impact on new jobs is some two to four times the numbers shown here, depending on specific assumptions and economic theories applied. Exactly how many of the new jobs are displacements of old jobs or creation of new ones is difficult to speculate. Very little displacement is anticipated since most of the new capabilities afforded by space industrialization are complementary to existing systems. . . . In the aggregate, the best guess is that 75 percent or more of the postulated space industrialization initiatives revenues will be job creating in the 1990s and beyond. Thus for a (total U.S.) work force of 100,000,000 in 2010, some three to twelve percent could be employed in new jobs created by space industrialization.

Table 2

New Jobs and Taxes Generated

New jobs (direct only, based only on U.S. markets for space industries.):

Year	No SPS	Baseline, with SPS	Upside
1985	15,000	100,000	120,000
2010	1,000,000	1,900,000	3,800,000

Taxes generated (direct only, based only on U.S. markets for space industries, 1977 dollars):

Year	No SPS	Baseline, with SPS	Upside
1985	$ 100 M	$ 800 M	$ 1,100 M
2010	$10,000 M	$20,000 M	$40,000 M

CHAPTER X
MANAGEMENT AND IMPLEMENTATION

FRAGMENTED ORGANIZATION
IN CURRENT SPACE PROGRAMS

Although we are now a quarter-century into the Space Age, thus far the U.S. has confined its military uses of space to support functions, such as communications, intelligence, and navigational aids. The fact that viable military defense options, based on known technology, had to be surfaced by High Frontier from outside the active military establishment strongly attests to serious conceptual shortfalls in that establishment. Indeed, there is a haunting parallel between the Pentagon's present perception of the military role in space, and the U.S. Army's attitude toward military aircraft missions in the early part of this century. Then, aviation was largely relegated to the Signal Corps!

There is no strong institutional voice within the DOD for the projection of U.S. military power into space. Space functions that *do* get attention—intelligence, communications, and navigational aids—are the responsibilities of a bewildering number of organizations. We will begin this chapter on management with a description of what High Frontier must avoid: the fragmentation of management and implementation. The following analysis of this fragmentation of our military space effort was provided by the office of Congressman Ken Kramer of Colorado, who has introduced legislation designed to accentuate the space mission of the Air Force, and to create a Space Command within that service. There was also support in and out of the Defense Department for a U.S. Space Force, with relationships to the Air Force akin to the Marine Corps's relationship with the Navy—or even as a new military service. Mr. Kramer's efforts were instrumental in causing the creation of an Air Force Space Command, now based at Colorado Springs. If the US Government should opt to implement the High Frontier Strategy, this Command would be keyed to success. Its critical job is to pull together the bits and pieces of the Air Force's *current* program.

The fragmentation of the American military space program was considerable by any measure. A brief look at who does what for whom in space, from the operational level up to the policy or planning level, exposes the complexity and organizational overlap of space activity. Of the three services, the Air Force operated with the greatest number of separate offices for space activity.

AIR FORCE

At least four major Air Force commands have been involved in space operations.

Strategic Air Command (SAC)

This command managed and operated the early warning and surveillance satellites and the ground radar systems that provided warning of Soviet missile attack. This activity was managed by the "SX" office at SAC headquarters. Another SAC organization, the 1st Strategic Aerospace Division, managed Vandenburg Air Force Base, the launch site for military Shuttle operations. The "1st STRAD" also ran the Defense Department's Defense Meterological Support Program (DMSP) weather satellites, and was in charge of the then-new navigation network—the 18-satellite Global Positioning System (GPS). Its planning activities for Shuttle operations were to be critical to the scheduling and turnaround of military Shuttle flights.

Air Force Systems Command (AFSC)

This command performed research, development, and acquisition for the Air Force of everything from armaments, radars, electronics, and space systems to new aircraft designs. AFSC has an internal Space Division in Los Angeles with a considerable number of program offices handling Air Force space programs. AFSC, however, usually ends up operating as well as developing these programs, even though it is not a fully operational command.

Because it had acquired an ad hoc operational responsibility as well as an R&D responsibility, AFSC established a new office: Deputy Commander for Space Operations (DCSO). The DCSO's primary responsibility has been running the Satellite Control Facility—a single facility controlling most Air Force and DOD satellites, as well as several Navy satellites. (Other Navy satellites are controlled from another facility in the U.S.)

Aerospace Defense Command (ADCOM)

This major command has provided many of the assets controlled by the U.S.-Canadian North American Aerospace Defense Command (NORAD). It also had a long-range planning staff at ADCOM headquarters that looks at space defense, antisatellite, and space surveillance operations. In addition, ADCOM/NORAD has operated the Cheyenne Mountain Complex, including the Space Defense Operations Center (SPADOC).

Air Force Communications Command (AFCC)

AFCC has run Air Force communications that are routed through space. The Air Force headquarters has also had several specialized planning offices that developed new missions or technologies for space. The Plans and Operations (Space) Office on the Air Staff was set up September 1, 1981. The group's establishment was the first formal Air Force recognition that the service had to develop an operational approach to space systems that matched what has long existed, as a standard planning perspective, for the aircraft side of that service. In short, its establishment represented an attempt to provide a centralized planning structure for space operations where none existed before.

Meanwhile, another separate headquarters staff, Research and Development (Space), conducted its own research activity on space systems. Coming from a research perspective, this organization interacted closely with specialized research organizations, such as Defense Advanced Research Projects Agency (DARPA).

The Deputy Assistant Secretary for Space Plans and Policy is under the Office of the Secretary of the Air Force. This individual is the Air Force's highest-ranking civilian official whose primary responsibility is formulating Air Force space policy.

The Air Force also participates in a tri-service planning effort that may be of considerable importance in the future. This program, called TENCAP (Tactical Exploitation of National Capabilities), is designed to extract useful tactical information from surveillance satellites and other sensors already operating in space.

Finally, there are three separate offices or organizations within the Air Force that plan or operate classified programs, including those conducted with other agencies or services.

NAVY

The Navy, meanwhile, has its own recently established Directorate of Space Systems which handles all Navy space activity, including classified programs, communications, and Navy participation in the TENCAP program described above.

ARMY

The Army's Ballistic Missile Defense System Command is developing a considerable "space focus" in its own right. It looks at the deployment of long-range antiballistic missiles (ABMs) that reach and intercept Soviet missiles far from the U.S. homeland. It is also developing an optical probe for attack assessment, which will be fired from American territory into any approaching target mass to provide last-minute verification and tracking of an attack. However, NORAD will have operational control of any deployed ABM system (as it did during the Safeguard ABM system's brief life in 1975), and also of the optical probe. Lastly, there is a classified program run by the Army Space Program Office.

OTHERS

DARPA is DOD's primary technical research organization. It is critically involved in advanced space systems research and, as a consequence, is an indirect player in the space policymaking process. In the past several years, DARPA has provided crucial advocacy for directed energy weapons used in antisatellite operations and ballistic missile defense.

The intelligence agencies (CIA, National Security Agency, Defense Intelligence Agency, and other offices) also plan and operate major space surveillance systems.

The Defense Communications Agency (DCA) is in charge of the department-wide communications, including those routed through space. DCA coordinates Joint Chiefs of Staff (JCS) operational requirements for communication systems that other organizations must follow in space communications design.

Several crucial offices are deeply involved in planning space systems at the higher policy and planning levels, under the Office of the Secretary of Defense and the assistant secretaries. One such office is the Deputy Undersecretary of Defense for Command, Control, Communications, and Intelligence. This office plans and oversees the electronic systems, including those in space, that control and communicate with our forces or provide the large quantities of intelligence data that come from around the world. There is also a Deputy Undersecretary of Defense for Research and Engineering (Strategic and Theater Nuclear Forces). This office works on strategic attack forces, including laser weapons and theater nuclear forces.

The senior official within the Department of Defense who is charged with formulating and coordinating space policy is the Deputy Director for Intelligence and Space Policy. He reports to an Undersecretary of Defense for Policy, who reports to the Secretary of Defense. Recently, a high-level department-wide committee was established to coordinate defense space policy and activity; the Defense Space Operations Committee, DSOC, which is chaired by the Secretary of the Air Force.

All of the above organizations, commands, agencies, and offices operate within the defense establishment and its associated intelligence activities.

NASA, the civilian space agency, also maintains a close relationship with the military space program since its Shuttles, upper stages, and other operating elements are often developed and funded cooperatively with DOD. For example, it was a large military payload and a military performance requirement that determined the size and payload lifting capacity of the Shuttle bay. It also determined the Shuttle's ability to fly considerable distances within the atmosphere after reentry—its so-called "cross range" performance.

The Executive branch of the government is officially in charge of overall national space policy. This policy is undertaken at White House level; it includes the National Security Council and the President's Science Advisor, who is supported by the Office of Science and Technology Policy. The present Science Advisor, Dr.

George Keyworth, is (or was recently) conducting a space policy review for the Reagan Administration.

Obviously, the Air Force is by far the service with the most pervasive and diffuse organization for space. It is also the service most in need of reorganization in its space efforts. A recent symposium paper by Dr. Charles W. Cook, the Air Force Deputy Assistant Secretary for Space Plans and Policy, described in grim detail what this organizational fragmentation has meant:

> . . . One of the most serious consequences of the wide distribution of responsibility for space operations (within the Air Force) is the absence of any centralized point within the Air Force for conducting long-range planning for space systems and support functions . . . Planning has been hampered by a lack of vision within the Air Staff and the OJCS (Office of the Joint Chiefs of Staff) for space operations. It has lagged the potential for using the benefits of space. (*Editor's note: This paper was written eight months before the establishment of the new Air Force office.*) . . . Space operations comprise a young but steadily growing mission area. The associated technology, doctrine, and policy are evolving as the potential of space as a medium of warfare becomes increasingly apparent. As our national space policy and doctrine mature, it would be a grave mistake to fragment their growth among several commands. . . . There is no significant effort underway to develop doctrine, plans, and requirements for controlling (space) weapons. Without an assignment of responsibility for operations planning within a

centralized organization, attempts to develop space-based weapons and supporting C^3I (Command, Control, Communications and Intelligence) programs are likely to take longer and be more expensive. (Dr. Charles W. Cook, "Organization for the Space Force of the Future", drafted January 21, 1981. Presented at the Air Force Academy's Military Space Doctrine Symposium, April 1-3, 1981; pp. 479, 481, and 484 of the symposium report, *The Great Frontier: Military Space Doctrine, Volume II*.)

Since the delivery of Dr. Cook's pessimistic appraisal, the Air Force has taken action to redress the planning deficiencies he described. Nevertheless, the U.S. military space program needs considerable organizational realignment as we move through the 1980s; and a number of basic organizational and policy decisions remain to be made.

IMPLEMENTING THE HIGH FRONTIER

In previous chapters we have said that the High Frontier requires development of several new civil and military systems as expeditiously as possible. Not only must we acquire the initial first-generation systems quickly and economically; they must be acquired in a way which sustains their priority nature, avoiding any sizeable near-term budget add-ons or fiscal-year "balloons".

The time needed to implement our proposed space capabilities, especially on the military side, is critical to the effort's overall ability to recover the margin of safety that the President has promised. A brisk pace is also crucial to ensure that the Soviet Union does not achieve the capability to deny the U.S. access to space—either for national security or for economic purposes.

It is quite clear that we cannot develop these systems in a timely way unless we streamline the overall management structure. The average of 13 or more years (!) it now takes the DOD to acquire major new weapon systems is unacceptably long, and enormously expensive. In the 1950s, strategic systems such as Atlas and Polaris were selected, developed, and made operational in four to six years. At that time, those systems involved more technological unknowns than do many of the illustrative systems discussed in this book.

Of the numerous studies during the past 10 years on how best to acquire new defense systems, two have had recent impact. These studies were the Defense Science Board 1977 Summer Study of the Acquisition Cycle, chaired by Dr. Richard D. DeLauer (now Undersecretary of Defense for Research and Engineering); and the April 10, 1981 "Memorandum on Improving the Acquisition Process" by Deputy Secretary of Defense Frank Carlucci. Both reports are in general agreement as to why the time span for acquiring new defense systems has more than doubled since the 1950s. They also agree on corrective measures.

Corrective actions in all cases have been recommended and efforts are underway to implement these actions. Furthermore, precedent exists for shorter acquisition cycles, since we continue to pursue them in the case of some intelligence systems and commercial programs. Unfortunately, despite the recognized need by the DOD to drastically reduce systems acquisition time, progress to date has been limited. It is imperative that we further reduce layerization, overregulation, and bureaucratic resistance to levels needed for High Frontier systems.

In addition, this major new initiative with its national security goals calls for a highly visible new management organization. Without such an organization, the effort would quickly acquire the image of merely one more service or departmental proposal to build a few new NASA and Defense space vehicles or weapons.

NASA does not suffer as much as Defense from such things as overlayered organizations; competition for funding between multiple systems; and the many regulations that now stretch out acquisition times. Nevertheless, special implementation measures will be needed to expedite the NASA-managed systems, as well as those managed by DOD. Therefore, if the earliest possible operational capabilities are to be achieved and a Presidential initiative identity maintained, we should select and acquire our first-generation High Frontier systems under special organizational and procedural arrangements. We will reduce the overall program costs as we reduce acquisition time. However, these special arrangements need only be such as to ensure that the process can benefit by the measures previously recommended in the DeLauer and Carlucci studies—though in the context of a national program, perhaps with international aspects.

SPECIFIC PROBLEMS

Over the past two decades, the factors most responsible for the overly long time it now takes to acquire a new advanced weapon or space system have become:

1. Front End Decisionmaking.

This is time devoted to stating a firm operational requirement, and to selecting specific systems for full-scale development. The decisions necessary to implement the High Frontier program can and should be made by the President, on the advice and recommendation of a committee of recognized scientists and systems acquisitions experts.

2. Inadequate Commitment to Acquisition From the Outset.

We must have funding and assignment of top priorities for full-scale acquisition, to preclude delays that result from having to compete for, or await funding in, latter phases. We can prevent this in the case of first-generation systems by centralized advocacy for multi-year funding and "fencing" of the appropriations obtained.

3. The Reticence in Recent Years to Maximize Concurrency in the Acquisition Process.

We have added years to important programs through:

• Failure to initiate procurement of long lead-time production items during the development and testing phase.

• Failure to authorize full production prior to "first flight".

• Failure to conduct joint technical, operational, and user evaluation testing.

4. Insistence on Too Detailed and Specific Performance Requirements or Operational Capabilities.

The first generation of essential new "cutting edge" space and point defense systems can benefit from cost- and time-saving innovations proposed by contractors, providing that we do not specify performance and operational capabilities too rigidly. Departments must write broad, flexible specifications stating only what is needed. They should require only "sufficient" performance for the first-generation High Frontier systems. Industry should also be allowed as much flexibility as possible in responding to requests for proposals. All too often, a government request for a proposal specifies unnecessary detail that may actually *prevent* a contractor from proposing his best design.

5. Overregulation of the Acquisition Process and Excessive Organizational Layers in the Review, Approval, and Decision Process.

Timely procurement and cost savings will require waivers from strict adherence to regulations of the Office of Management and Budget, and of DOD. These regulations should be used only as guidelines. The chain of command for review and decisions for High Frontier systems, in all departments, should bypass all levels not in a position to make authoritative decisions on the issues concerned. To avoid delays, we must set up a streamlined channel of decision, review, and procurement, with High Frontier expediting offices at appropriate levels.

6. Personnel Selection and Motivation.

We will lose time and effectiveness unless we assign experienced, dedicated program managers and key people to the task at all levels. They should remain with the program throughout the acquisition cycle (four to seven years). Departments should take appropriate steps to meet these requirements, including provisions for promotion and protection from any loss of career opportunity. We should make similar provisions for key personnel in the High Frontier centralized organizations.

7. Distrust and Misuse of Contractors.

The complexity of advanced new systems, the frequent need for decisions on changes or alternative approaches to resolve technical problems, and the ability to exploit money-saving contractor suggestions all require close teamwork. This means close cooperation between the contractors, the program managers, and the eventual user organizations. In recent years, a distrust of contractor motivations and/or fear of adverse publicity and criticism resulting from close personal relationships between government officials and contractors have worked to the detriment of team efforts. Yet this teamwork is absolutely essential to efficient, economical acquisition.

8. Excessive Micromanagement.

Appropriate decisions should be made at those levels where the expertise exists, to prevent wasted time and costly mistakes. In recent years, growing congressional staffs have too often resulted in young, dedicated but inexperienced staffers seeking to influence technical and operational decisions that should not be dealt with at the congressional level. The tendency toward micromanagement of systems acquisition programs by congressional or DOD staffers, and by principals at levels not directly responsible, has resulted in delays, waste, and the risk of bad decisions.

ALTERNATIVE ORGANIZATIONAL APPROACHES

We have examined three basic alternative approaches for acquiring the first-generation High Frontier systems. These were:

- To establish a new, separate, centralized organization to manage all aspects of all High Frontier systems, by incorporating the initiatives to streamline the system acquisition process.

- To establish a special task force to select the desired systems, and then assign the entire acquisition task to the department having primary interest in each system.

- To establish a centralized, interdepartmental organization which would not only select the specific systems to be acquired for the President, but would also follow through on their acquisition by the responsible departments, by retaining certain continuing policy, funding, public information, and other responsibilities.

In brief, the choice was to do the systems selection, development, and procurement tasks, and to defend and justify the programs under a new organization in toto, under existing departments, or under some combination of the foregoing.

Separate Centralized Organization

Acquisition under a separate, new, centralized organization would ensure maximum exploitation of all shortcuts to include quick, decisive selection; effective funding; outstanding top management; clear priority, immediate decisions when needed; and freedom from in-house competition, turf-guarding opposition, and military service or other special-interest influences. Such an organization would also best accommodate the interdepartmental and, if so decided, the international nature of High Frontier.

The government's resources, and the administrative and research support required for any major high-technology systems, now lie mostly within the Department of Defense or NASA. Duplicating these, or transferring them to a new, centralized organization, would be time-consuming and costly. It undoubtedly would meet strong resistance within the bureaucracy; and it would suffer from disinterest or refusal to serve on the part of many experts whose careers are tied to their departments or to military services. Major legislative problems would also be created in connection with the authorizing, funding, and manning of new, separate organizations whose mission would clearly conflict with that of both DOD and NASA.

On balance it appears that any savings in time, or enhancement of the effort's image, resulting from this centralized management in a new organization would be more than offset by the problems it would generate. We would almost surely encounter personnel, administrative, legislative, and morale problems, as well as delays in trying to resolve these problems.

DELEGATION BY DEPARTMENT

The second alternative considered was selecting the desired systems and then delegating all High Frontier implementing activities to the departments. The President would approve the concept; appoint a Systems Selection Task Force to pick the the first-generation systems to be acquired; and then disband the task force and direct the existing departments to fund, defend, and acquire the chosen systems as a matter of priority.

This alternative is the simplest solution *if* (1) the departments can be depended on to acquire new systems expeditiously without continual central supervision, and (2) some provision could be made for the high visibility, interdepartmental—perhaps international—nature of the effort.

Unfortunately, to judge from DOD acquisition statistics and related studies, the departments have been unable to implement all the measures they themselves recognize as essential for rapid, cost-effective system acquisition. Under normal departmental management, High Frontier would quickly lose its character as a "bold new Reagan initiative". Its systems would soon be reduced to "just another program competing for funds and management skills," regardless of both Presidential and Secretary-level directives. This general malaise results from a combination of:

- The large size of the departments.

- The many organizational layers, especially in DOD.

- The inherent compromising in the joint service structure.

- Inevitable turf-guarding of programs that will suffer from High Frontier priority funding.

- The requirement to operate within existing policies, procedures, and regulations.

- Interservice rivalries.

INTERDEPARTMENTAL ORGANIZATION

In view of the shortcomings inherent in these first two alternatives, the third alternative—assigning acquisition responsibilities to appropriate departments, with selected responsibilities for certain aspects of the program delegated to new, centralized organizations—appears to be the best means of meeting all the vital requirements of the High Frontier initiative.

The urgency of deploying first-generation systems to cope with the Soviet threat, and the need to overcome delays inherent in departmental acquisition efforts, call for separate, top-level, overall direction of the effort. Concurrently, the fact that the departments have the majority of the U.S. facilities, expertise, and personnel resources required for such a task calls for their being assigned that task.

The solution that appears best to achieve the desired goals, and to meet legitimate objections, would be one in which:

- Several centralized organizations would be assigned such tasks as overall policy direction; specific system selection; acquisition and allocation of funds; granting of priorities; rapid decisionmaking where necessary above departmental level; and public and congressional relations.

- Specific system-acquisition management, however, would be assigned to the interested departments.

Under this solution, the centralized organizations would consist of: (1) a National Space Council under the chairmanship of the Vice President; (2) a Systems Selection Task Force (SSTF) reporting directly to the President; and (3) a Space Consolidated Program Office (SCPO) also reporting to the President but through the Chairman of the Council.

These organizations would be established and defined by Executive Order. NASA would be designated the executive agent for housekeeping for the new, centralized offices. Consideration should also be given to having the deputy administrator of NASA as head of the SCPO.

Specific systems for initial procurement would then be selected and recommended to the President for urgent acquisition by the SSTF. Executive Orders, prepared by the SCPO for the President, would direct the responsible organizations to acquire systems selected as a matter of priority. Specific authorizations, exceptions to regulations, priorities, and appropriate Presidential guidance, etc., could accompany these Executive Orders as necessary. This would ensure that specific system-acquisition management personnel have the freedom and support to vary from current acquisition procedures whenever this could save time or money.

The National Space Council would oversee the High Frontier initiative and the activities of the SCPO. In addition, it would provide a source of quick and final decisions, when and where needed, during the acquisition process.

SYSTEM SELECTION PROCESS

The DOD studies of acquisition processes, to which we referred earlier, indicate that great time savings can be obtained by minimizing the time it now takes to agree on the operational requirements, and on specific systems to be developed to meet these decisions. These are the "front-end" decisions for any new program.

The period from the identification of a requirement until full system development is approved (historically called by several different names: "conceptual phase", "phase O", or "definition phase") has steadily increased over the past two decades. Today it frequently exceeds *six years*. This can be attributed to a lack of decisiveness from top leadership as to what is wanted; and the substitution—instead of a decision—of excessive studies, debates, analyses, and reviews. The current approach to defining requirements, and to defining what will be built to meet these requirements, would introduce unacceptable delays in deploying High Frontier systems.

From a review of the overall state of the art and of many industry proposals, it is clear that the desired systems can now be expeditiously selected, built, and deployed; and given adequate priority, funding, and special management arrangements. The President can make these decisions on the advice of the SSTF in lieu of waiting for voluminous, formal industry or departmental proposals. When and if made, the President's decisions should be accompanied by a commitment to procurement from the outset. These Presidential front-end decisions could save up to six years in acquiring High Frontier systems.

In the 1950s a situation existed similar to what we face today. At that time, the Soviet threat called for the development of U.S. intercontinental ballistic missiles, but there were competing DOD requirements and industry proposals. As a result DOD, and specifically the U.S. Air Force, lacked the motivation and/or technological confidence to direct full-scale development and procurement of first-generation ICBM system concepts. President Eisenhower then appointed the Von Neumann Committee, consisting of recognized scientific and defense authorities, to select the specific ICBMs to be built at that time. On the basis of their advice, he then directed the development and procurement of three systems: Atlas, Titan, and Thor.

Our recommendation—that President Reagan establish a High Frontier SSTF to review various technical, service, and industry proposals for first-generation High Frontier systems—was based largely on this historic, successful approach which quickly initiated a new national high-technology program. Such a SSTF would be a one-time, ad hoc review and selection board which would report directly to the President (see Figures 32 and 33). When the operational requirements are established by approval of the High Frontier proposal, the job of the SSTF would be solely to select and recommend specific systems to the President. The departments should be immediately directed to acquire those systems. The SSTF should be made up of leading U.S. aerospace scientists, industrialists, and management experts with such other technical expert members as the President might desire. We estimate a membership of between 9 and 12 individuals, with staff support provided by the SCPO as discussed below.

The role of the SSTF would not be to debate the Assured Survival concept, or the nature of this initiative, but *only* to recommend systems that can be quickly obtained to achieve the operational capabilities called for. The SSTF should be requested to render its recommendations within four months of activation, and would then be disbanded.

National Space Council

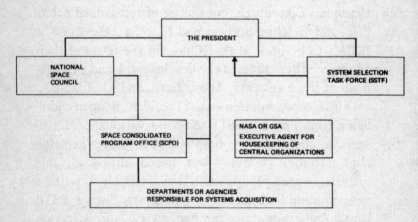

Figure 32. High Frontier Recommended Implementing Management Organization

In order to ensure coordinated direction for the High Frontier initiative at the highest level, and to allay any concerns within the Executive Branch that the centralized organizations would usurp their responsi-

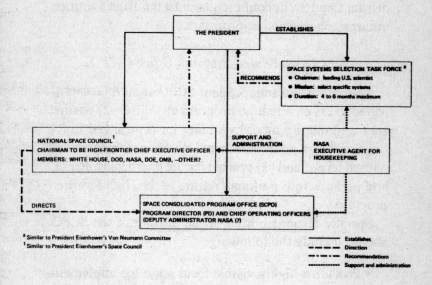

Figure 33. Proposed High Frontier Acquisition Organization — Top Echelon

bilities, we propose a National Space Council somewhat similar to the one established under President Eisenhower in 1958. The Council would oversee the implementation of the High Frontier effort on behalf of the President. It would provide broad policy guidance to the departments involved and would direct the activities of the SCPO. The chairman of the Council would be the Vice President. He would act as the chief executive officer of the High Frontier program as well as its principal liaison with the leaders of Congress in seeking their support and the funding for the program.

The Eisenhower-era council performed a useful role when the Soviet Sputnik event provided both a threat and an opportunity. Today we again face both a threat and an opportunity in space; hence, a similar Presidential Council, in conjunction with the High Frontier initiative, is deemed appropriate.

Space Consolidated Program Office (SCPO)

A separate and independent SCPO should be established to (1) ensure high program visibility; (2) ensure staff and management capability for centralized functions and as a source for rapid decisionmaking at the highest levels; and (3) provide for the interdepartmental, and perhaps international, nature of the High Frontier program.

Specific responsibilities of the director of the SCPO should include the following:

- Provide a highly visible focal point for implementation of the Reagan High Frontier initiative; and act as secretariat to the Council and SSTF.

- Draft Executive Orders and guidelines for departmental acquisition of selected High Frontier systems.

- Seek funding from Congress; and justify, obtain, defend, and allocate this funding for the selected systems.

- Seek and obtain Presidential, Office of Management and Budget (OMB), or Congressional waivers of laws or regulations as necessary for departments to expedite acquisition.

- Work with the OMB and Congress to reprogram funds among the various High Frontier system program offices where and when desirable.

- Resolve interdepartmental issues or refer these to the National Space Council, as appropriate.

- Provide or seek from the Council any decisions requested by the departments to expedite the program.

- Serve on any systems acquisition and review committees established by the departments to oversee High Frontier programs.

- Install and operate a central management information system to monitor High Frontier programs; and provide the President and the Council with frequent status reports.

- Coordinate with the Department of State any international negotiations relating to the High Frontier.

Departmental Roles

The departments will be assigned specific responsibility to develop and acquire specific High Frontier systems as directed by the President. This will be done by Executive Orders directing acquisition of each system. The departments should be requested to set up special High Frontier offices at appropriate levels to expedite decisions. Direct communications should also be authorized with the centralized organizations as needed to achieve these objectives. The goal should be to organize the effort to eliminate from the chain of command and review, any organization or staff level that is not specifically assigned High Frontier acquisition responsibility, and/or final decision authority.

The Executive Orders to acquire specific systems might specify the following roles for the responsible departments:

- Undertake the full-scale development, test, evaluation, procurement, and operation of designated systems.

- Accomplish the foregoing in accordance with the priorities and policies assigned by the President and Council, and by setting up special system program offices (SPOs) to staff these tasks with outstanding people.

- Adopt special management procedures and communication channels, and delegate authority as necessary, to permit High Frontier systems to benefit from all recommended acquisition actions in recent studies.

- Cooperate with the SCPO to establish direct communication channels, and to implement a centralized management information system.

- Ensure that the funds appropriated for High Frontier acquisition are "fenced" and disbursed per the National Space Council's policies and directives.

- Support the National Space Council and the SCPO in their assigned missions.

The above department guidelines are only illustrative examples. Actual guidelines would be discussed with department secretaries and reviewed by the Council before inclusion in Executive Orders. These orders would be prepared at the time when responsibility for each system is assigned, since orders will differ between departments and between systems.

SUMMARY: PREREQUISITES FOR RAPID, EFFICIENT ACQUISITION

KEY INITIAL ACTIONS

1. The President assigns the highest priority to a space consolidation program.

2. The President announces the urgent need for the High Frontier program, and periodically explains to the nation how an extraordinary effort in space will provide our margin of safety.

3. The President announces the appointment of a Systems Selection Task Force (SSTF) with a finite life, like the Von Neumann Committee of the 1950s, composed of recognized scientific, defense, and acquisition authorities. Its purpose is to review candidate space systems and recommend to the President, within a few months, the best systems to pursue.

4. The President announces the appointment of a National Space Council with a finite life span, consisting of high-level membership from the White House, National Security Council, NASA, DOD, and others. Its purpose is to oversee the progress of acquiring selected systems, and to resolve Congressional and interdepartmental issues.

5. The President announces the establishment of a Space Consolidated Program Office (SCPO), to provide centralized integration and direction for High Frontier programs.

6. The President instructs all government agencies involved in this initiative to use realistic inflation factors in cost planning.

ORGANIZING

1. The President issues Executive Orders and Presidential Directives establishing the SSTF, the National Space Council, and the SCPO.

2. The President selects membership for the Council and SSTF, and issues an Executive Directive to NASA identifying NASA as the executive agent for housekeeping and logistic support of the new, centralized organization.

3. The Council chairman (the Vice President) selects an SCPO director of national stature and experience, who should report directly to him. The program office will also serve as secretariat to the Space Council and to the SSTF in addition to its other functions.

4. Departments assigned responsibilities for acquisition establish system program offices (SPOs) to acquire the selected systems. They assign High Frontier expediters at departmental and systems-command levels.

5. The SCPO director determines the type of systems, engineering, and technical assistance support required by their SPOs, traditionally provided by—though not limited to—Federal Contract Research Centers (FCRCs). These support groups should be under the command and control of SPO managers.

6. Departments direct government representatives at major contractor plants to be responsive to appropriate system managers on High Frontier programs.

7. Liaison personnel, from operational users and from major contractors, are assigned and collocated at the SPOs.

8. The High Frontier Systems Acquisition and Review Committee is established to include the SCPO director as chairman, with members from responsible department head offices, system commands, or NASA centers; and the interested systems manager is established. This is to furnish the DOD Defense System Acquisition and Review Committee (DSARC) role, and also to provide quick decisions when needed.

PERSONNEL CONTINUITY

1. The departments select and assign the best available talent in their departments to the SPOs.

2. All consolidated program and SPO personnel are assigned for the duration of their respective systems (approximately four to seven years), and in manning positions above their current grade to provide for in-place promotion.

DECISIONMAKING PROCESS

The overall process is reflected in Figure 33. The following details are steps in that process.

1. The President and Council seek an agreement with Congressional leadership to rely on the full

reporting to (and monitoring by) the SCPO and Council. This is in lieu of the prevalent micro-management by Congressional staff and subcommittees which causes program delays.

2. The departments are authorized to place program objective memoranda and budget cycles on a multiyear basis, to be forwarded by the Council's chief executive officer to Congress for disposition.

3. Authorization is granted to departments to delete determination and findings from the budget process. The SCPO director works out a streamlined spending authorization mechanism to be approved by the Council chairman and by Congress as necessary.

4. Congress is persuaded to provide multiyear funding which will be "fenced" at department level (a procedure already used in intelligence programs). Congress provides needed reprogramming authority for the SCPO director to include additional funds for the systems office to handle uncertainties and unforeseen technical problems.

5. Changes to overall system performance not resolved by the Acquisition Review Committee are referred to the Council chairman who acts on advice of the SCPO program director. As a ground rule, all levels consider performance requirements "sacred" or unalterable, if significant time and costs can be saved while providing "sufficient" system performance to obtain an Assured Survival capability.

6. First national priority is assigned for critical, exotic, limited availability and long lead-time materi-

als which are required for for the development, test, or production of selected systems.

7. The Council and departments delegate maximum authority and flexibility, as proposed in all recent acquisition studies; grant direct communication with any and all government decision levels; and free system managers from having to consult any official who cannot provide decisions.

8. Approval of the SSTF recommendations constitute Milestone Zero. Those systems then approved for acquisition by the President go directly into DSARC II — full-scale development phase — by Executive Order with full intent to deploy these systems.

9. Concurrency is the rule consistent with risk throughout the acquisition cycle. This is especially true between full-scale development and production; initiating and developing ground, logistic, training, and personnel support; command, communication, and control subsystems; and system facilities. Production is continued prior to completing the operational test and evaluation phase, and joint user/ developer/contractor tests are conducted with full evaluation.

10. Departments are granted authority to waive existing DOD and OMB directives and regulations for acquisition management; requests for proposals; statements of work; and system, subsystem, and end-item requirement specifications. Directors' and managers' judgments are used in accord with the spirit of the directives and regulations, rather than applying a literal or stringent interpretation.

11. SCPO director's and system manager's discretion is applied for fair, quick, and flexible source selection procedures, with approval as needed from the Council chairman. The same flexibility applies to the form of contracts used for each system, where all types are considered and carefully selected to best match the needs of each particular item and phase of the acquisition cycle.

OVERALL IMPLEMENTATION SUMMARY

The basic implementing plan, then, for first-generation High Frontier systems is to provide for special, centralized policy and management only as deemed necessary to meet overall goals. These goals are:

- Select the systems to be acquired.

- Maintain the identity and priority of the "initiative".

- Deal at the highest level with the funding justification and allocation, and with overregulation problems.

- Ensure quick decisionmaking and risk-taking where appropriate.

The acquisition task for each specific system would be assigned to the responsible department (presumably, but not exclusively, NASA and Defense) by Executive Order. This order should include Presidential instructions to the departments to take organizational and procedural measures necessary to minimize delays and to achieve operational capabilities at the earliest possible dates.

We have proposed a centralized Executive-level organization, consisting of three separate entities (see Figure 32). These would be established by Executive Order as previously noted, and would exist only as long as necessary to acquire and deploy the first-generation systems. Housekeeping should be assigned to NASA. These three temporary organizations are essential to expedite the systems, sustain priorities, and oversee implementation of initiatives to improve the systems acquisition process. Responsibility for acquiring each system should be assigned by Executive Order to the department having primary interest. The Executive Order should contain special procedures and organizational relationships, to eliminate major delays in the acquisition process.

Presidential instructions to adopt some or all of these measures should not be deemed an infringement on departmental responsibilities. Bypassing established organizations such as the Joint Chiefs of Staff or military services can be done with Presidential authority. Any such instructions would be routinely reviewed by the proposed National Space Council, where affected departments and agencies would be represented. Insofar as the DOD is concerned, the proposed special management measures are supported by the findings and recommendations of the DOD acquisition studies, such as the DeLauer and Carlucci reports, and current DOD policy documents on systems acquisition (DODD 5000.1 and DODDI 5000.2).

CHAPTER XI
CRITICAL VIEWS

Any major shift in U.S. national strategy is bound to garner criticism, and the High Frontier is no exception. A continuing dialogue has already surfaced in the media between proponents and critics of the High Frontier. As of March 1983, the Soviets have made no explicit response; but given their oft-stated views, we can readily predict a flood of opposition, as we will describe below.

DOMESTIC CRITICS

We may expect criticism from well-meaning analysts who argue against any kind of military presence in space. While clearly aware of orbital Soviet weapons, some will still argue for "unilateral restraint" on our part. Thus Daniel Deudney, of Worldwatch Institute, writes that U.S. development of ASAT weapons would provoke a new arena of arms competition, adding, "Unilateral U.S. restraint and an immediate resumption of negotiations would do more than further tests to enhance the security of both nations." He does not dwell on the fact that, given the existing Soviet ASAT capability, our restraint would amount to the kind of appeasement which England's Neville Chamberlain offered Germany until 1939, with disastrous results. Deudney states that an orbital beam weapon system against ICBMs " . . . would cost—conservatively— between a phenomenal $500 billion and $1 trillion. . . . " We might expect similar extraordinary figures to accompany criticisms of High Frontier programs. While claiming exorbitant costs, critics may ignore the tremendous *returns* on those costs when space industry begins to expand on the High Frontier.

Some critical appraisals of point defense systems also include much higher cost estimates than we estimate for some candidate systems. Kosta Tsipis of MIT cited two objections to point defense in the *New York Times*, December 1, 1982, in this short passage:

> The only way to reduce the vulnerability of the (MX) dense pack would be to place an antiballistic missile system around it. But that would require abrogation or renegotiation of the 1972 ABM treaty with the Soviet Union and would probably cost an additional $25 billion, doubling the price of the Dense Pack system.

Some proposed ABM systems would, no doubt, involve such costs; but we have described at least one point defense system in Chapter VI (the GAU-8 gun system) which would cost far less, *and* which does not violate the ABM treaty. In any case, our purpose is not to defend a specific point defense proposal; our purpose is to show that the U.S. has the technology from which we can select more effective—and that means cost-effective—systems.

Dr. Eberhardt Rechtin of The Aerospace Corporation levied modest criticism of High Frontier in *Military*

Electronics/Countermeasures magazine in November, 1982. Rechtin believes that the hostile warhead "leakage" rate through illustrative High Frontier GBMD space-based systems, as well as the costs, would be higher than those of groundbased BMD systems. This leakage, Rechtin said, could be diminished enough to protect our own deterrent missiles but probably would not save most of our cities.

The High Frontier team agrees that some leakage must be expected; but there are ways to minimize it. For two examples: decoys and stealth technology could be added to spacebased GBMD layers to improve their survivability with a relatively small exertion on our part. The result would be to enormously compli- cate the imponderables to a Soviet first-strike planner— which is a primary element of deterrence.

In the same article, Rechtin specifically endorsed the goal of the High Frontier:

The real importance of the (High Frontier) pro- posal seems to be that people should be paying special attention to this problem, that we should start to focus once again on ballistic missile defense. I think there is an increasing apprecia- tion that if you are attempting to make sure your deterrent will survive in the long run, some form of active defense is likely to be needed in the future. This (High Frontier) is one proposal.

To this we must add that ours is one proposal which would also place U.S. industry in space for peaceful, economically sound purposes.

The various groups which support bans on nuclear weapons—or on peaceful nuclear technology as well—

may not at first perceive that the High Frontier approach can yield much or all that they ask. With nonnuclear defensive protection for our deterrent forces, we need not further add to our nuclear deterrents; indeed, we might sensibly reduce their number. In addition, public fears of nuclear power plants can be resolved, even with apparently contrary public demands for cheap power, when solar power satellites can provide safe power from orbit. Perhaps the main continuing opposition to High Frontier from these quarters will come from genuine Luddites who place a blanket of disapproval over all technological advances.

Religious controversy will surely touch the High Frontier concepts. A draft pastoral letter circulated among U.S. Catholic bishops in 1982 categorically opposes U.S. weapons development. Its effect would clearly be to perpetuate the fallacies of Mutual Assured Destruction—including the one-sidedness of that destruction when the Soviet populace has its present Civil Defense advantages. Perhaps the best answer to religious criticisms of the High Frontier is that we do *not* object to a similar program by the Soviets. If neither antagonist can mount a successful first strike, we can all breathe more easily in a stable situation which does not hold our entire population hostage, as MAD does.

THE SOVIET VIEW

In the Kremlin's perception, the "big question" about the High Frontier decision will not be whether we have the technical, industrial, or managerial resources to carry it through. The big question will be whether

U.S. leaders and the American people have the will to do what is necessary to carry this nation's technological and industrial might to the High Frontier. We should not, however, expect to hear this question aired prominently by the Soviets in their criticisms. To understand the criticisms we *will* hear, one must first understand the frustration radiating from Moscow of late.

Moscow, over the past decade, has increasingly convinced itself that the U.S. is in a state of steady decline, both domestically and internationally. A variety of factors is offered by the Soviets to explain this decline. These hypothetical factors rest on the panoply of stock-in-trade Marxist/Leninist theories. They relate to the workings of "immutable" laws of social development; and they dictate both the fall of capitalism and the "inevitable" global triumph of Soviet communism.

The Soviets have emphasized a succession of concrete examples in which the U.S. has exposed its lack of will to "act in old ways", and our failure to confront the USSR from "positions of strength". For examples:

- The acceptance of defeat in Vietnam.

- The concession of strategic parity to the USSR, first in practice in the late 1960s and then formally in 1972.

- Acceptance of the Soviet-structured "principles of peaceful coexistence" in U.S./Soviet relationships since 1972.

- Repeated backdowns in the 1970s when U.S. and Soviet interests clashed at different points in the world.

• Consistent retrenchment in our military application of scientific and technological potential.

By the end of 1980, the Kremlin appeared to genuinely believe that the U.S. had reached a point of no return in its decline. With this irreversible shift in the "correlation of forces" in their favor, the U.S. could no longer choose its courses of action, but must act in accord with "objective realities" of superior Soviet power.

How does the Kremlin see these prospects in light of the Reagan Administration? Does it consider him bound by immutable limitations? Or does it see him as defiant of objective realities (as the Soviets define them), willing and able to follow follow courses he sees necessary?

The answer is central to the more specific question of how seriously the Kremlin will view a High Frontier decision by the President. The Soviets have come to react with great skepticism to deliberately arranged revelations of new U.S. approaches in military affairs, thanks to their standard appraisal of us. A prominent Soviet Washington-watcher has said, "It is very typical of American political tradition for new leaders to formulate some kind of new policy line, give it an eye-catching title, and thus 'declare' themselves without, however, any readiness to do the things necessary to effect any meaningful change." So Soviet military analysts lightly disregard such past-stated U.S. doctrines as "massive retaliation", "city avoidance", "flexible options", etc. To them, these phrases represent nothing more than empty subterfuges to suggest capabilities that we do not have. During the last months of the

Carter Administration, Soviet observers openly disdained U.S. announcements and contrived leaks regarding new weapons, strategies, and programs. Thus it makes sense to ask whether the Soviet perceptions have changed during this Administration.

At present, Moscow gives convincing evidence that it sees a genuine turnabout in U.S. international attitudes; and that it sees a formidable challenge of indefinite duration across the broad spectrum of "global communist struggle". Moscow indicates that it finds our present policies and actions decidedly disturbing, and not in line with their expectations. It sees a different ballgame from the one in which it has been engaged for the past 20 years. It feels that it is no longer testing us, but is being tested by us. In short, Reagan policies have profoundly shaken the solid Soviet conviction that his pre-election rhetoric would crumble in the face of world power "realities"; that he would perforce follow the same policies toward the USSR as his three predecessors.

During his visit to Moscow in 1972, President Nixon surprised the Soviets by accepting their principles of "peaceful coexistence" and by welcoming Soviet strategic parity. From that time until very recently, the Kremlin believed that U.S. leadership made this shift not through a change of heart, but by recognition of a change in the "correlation of world forces"—primarily with the military balance favoring the USSR. This belief hardened into conviction.

By the time Reagan assumed our Presidency, Moscow viewed the future as its oyster. Reagan might prove troublesome at first, but as soon as he saw the "actual" balance of world power, Reagan—like his predecessors

—would be forced to seek accommodation with the USSR.

Brezhnev had already assured a Central Committee Plenum on the eve of the 1976 election that "whoever might be in office in Washington . . . the United States will have to reckon with the actual alignment of forces in the world which in recent years have prompted the American ruling circles, after making a sober analysis of the existing situation, to seek ways of coming to an understanding with the world of socialism." The day after Reagan's election, Alexandr Bovin, a leading Soviet commentator, asserted that the same "realities" would guide Reagan.

Soon, Soviet spokesmen were nervously insisting that, as *Pravda* put it on March 22, 1981, "The captains of Western policy will sooner or later have to heed the inexorable realities of our time. . . . No one is going to turn back the wheel of history."

Eventually Moscow began to accept the fact that the Reagan Administration was showing an unexpected imperiousness to limits on its freedom of choice; limits supposedly set by the "objective realities" of an adverse "correlation of world forces". Politburo member and Defense Minister Dmitri Ustinov, at the highest level of Soviet authority, released a truly remarkable article made prominent in *Pravda* on July 25. Ustinov detailed the benefits to the Soviets from international relations in the 1970s. But, he lamented:

At the turning point of the 80s there has been a radical change in the policy of the United States and a number of other NATO countries. The upper hand there was resumed by circles which orient themselves on force in international matters, which refuse to accept the changes conditioned by history which are taking place in the world, and which have set for themselves the objective of altering, in their favor at any price, the balance of power in the world arena.

According to its own testimony, what is most disturbing to the Kremlin is that the U.S. Administration is evolving a comprehensive international policy and program of action which treats the Soviet threat as central and, on the other hand, avoids direct engagement of the U.S. with the USSR. Moscow has long been accustomed to our preoccupation with *relations* with the USSR. This is the first time since the 1950s that it has had to deal with a preoccupation with *opposition* to the USSR. The Soviets make it clear that Reagan's policies are having a telling effect on the Kremlin's assurance that it is in the driver's seat, knows where it is going, and by what means. We are now seeing signs of uncertainty, if not confusion, and a seeming loss of a sense of direction in the Kremlin. Soviet spokesmen leave no doubt that the Kremlin now regards the U.S. Administration as fully resolute in its purposes and knowing in its plans; hence both intent on, and capable of, major efforts to reverse the roles in U.S./Soviet relationships.

Several factors are evidently responsible for this new situation:

1. The Administration's Pervasive Focus on the Soviet Threat. Soviet spokesmen constantly speak of present U.S. policies as being taken up with "anti-

Sovietism", "anticommunism", and "the Soviet threat". The Soviets themselves have, of course, *always* stressed—indeed, made a virtue of—their own focus on the antiimperialist (i.e., anti-U.S.) struggle in virtually everything they do or plan to do. They do not like finding the shoe on the other foot. Also, unlike the previous administration, Reagan and his advisors have shown that they do not intend to be influenced by concerns over displeasing Moscow. Cases in point have been the Administration's stand on the Cuban/Soviet involvement in El Salvador; our open opposition to the Soviet threat to Poland; and our disregard of Soviet sensibilities in our moves toward closer relations with China. There is also the matter of calling a spade a spade. Something of a shock wave has been produced on the Soviets by the new Administration's practice of speaking frankly about the motives, behavior, and prospects of Soviet leadership. When President Reagan spoke at his January 29, 1981, press conference of Kremlin aims to establish a world communist order and made reference to the peculiar Soviet rules of morality, Moscow vented screams of rage. So, too, when former Secretary of State Haig charged the Soviets with using international terrorism in their conduct of foreign policy. By their reactions, the Soviets make clear that they perceive they are losing a one-way advantage they have long enjoyed in conducting ideological warfare against the U.S.—more than this, by a United States no longer willing to tie its hands when engaged in a struggle for survival. Clearly, this troubles Moscow no end.

2. Downplay of Dialogue and Negotiations. The Reagan Administration is proceeding with its international programs without directly engaging the USSR, despite mammoth Soviet efforts to involve us in various negotiation tracks. Brezhnev personally served up a heavy menu of proposed talks, agreements, and negotiations to the Reagan Administration. Its initial purpose was to involve us once again in the same sort of gamesmanship that had proved so useful in forestalling U.S. initiatives before. None of the proposals were new or involved any change in long-held Soviet positions. All were designed to put responsibility for concrete responses on Washington; to force us either to get involved with the USSR on their terms or, if we did not respond or responded negatively, to expose the U.S. as recalcitrant aggressors in contrast to the peace-loving Soviets. This has all been well-explained by the Soviets themselves. Central Committee Secretary Boris Ponomarev wrote in *Kommunist* in March, 1981, "The USSR's peace proposals raise a dilemma for American ruling circles: either to positively approach the peace proposals or to demonstrate their disregard for people's aspirations, and thus be viewed as warmongers." However, the Reagan Administration proceeded to develop its international program without responding to Soviet efforts to involve us in various negotiating tracks.

3. A "Global U.S. Offensive" to Reestablish Positions of Strength Worldwide. There is often an evident Kremlin perception that in lieu of seeking agreements with the USSR, the present U.S. Ad-

ministration follows its *own* agenda to develop positions of strength wherever U.S. and Soviet interests clash. Moreover, the Kremlin accurately perceives that we are having some success at it. *Pravda* reflected this view on March 25, 1981 asserting that "the White House can find nothing better than to settle the most complicated and acute problems of international life by means of arms and positions of strength." (As compared, for example, to the delicate Soviet finesse in treating its problem with Afghanistan.) A month later the leading Soviet foreign affairs journal, *International Affairs*, stated that what disturbs Moscow is that it sees the Washington government systematically undertaking to redress the shift in the correlation of world forces—which the Soviets had viewed for years as having moved irrevocably in their own favor.

4. The Reagan Commitment to Redress the Military Balance. Linked to all Soviet concerns about Reagan policies is the Kremlin perception that the President intends an unrelenting effort to refurbish U.S. military power as an indispensable prerequisite for an effective policy toward the USSR. This suggests a new arms race challenge to the USSR and, more important, a new challenge to the whole body of Soviet global objectives and expectations. Brezhnev said in a major speech in Kiev on May 9, 1981: "But there are also such statesmen in the bourgeois world who, judging by everything, are accustomed to thinking only in terms of strength and diktat. They actually regard the attainment of military superiority over the Soviet Union as their

main political credo." Other top Soviet authorities, particularly military, have not only echoed Brezhnev's charges about the new U.S. military efforts; they have added strong notes of alarm about the warlike purposes of the effort. Defense Minister Ustinov, in a July 25,1981 *Pravda* article, capped these statements of concern by citing figures on projected U.S. defense spending and asking why the U.S. needs such high expenditures. In answering his own question, Ustinov revealed one of the major Kremlin worries: that the U.S. will break out of deterrence constraints; that we will position ourselves to use force to protect our interests at any point on—or over—the globe.

Ustinov's answer, in part: "This whole course of the White House, which is dangerous for the cause of peace, is aimed at giving itself an excuse to react to possible conflicts occurring in any part of the world by means of military force. Not since the days of the cold war has the line of using force shown itself so plainly in U.S. policy. 'We must restore the high mobility of the armed services,' Secretary of Defense C. Weinberger asserts, 'and react quickly to changing situations in any part of the world—we must strengthen our positions in the world by means of arms....'"

Judging from these Soviet perceptions, the Kremlin appears to be readying itself for a trying struggle with the U.S., both for the short and long hauls. Apparently it no longer expects to ride the momentum of past successes to more decisive gains over a retreating foe which has little confidence in its ability to halt that retreat.

Still, Moscow professes to believe it enjoys important advantages; that these advantages can be extended; and that the USSR can aggressively continue to advance across a global spectrum. It clearly prefers to seek the revival of a Soviet-style detente relationship with the U.S. Then it would be able to press forward toward its own goals while the U.S. subjected itself to new self-imposed restraints in the interest of a hopelessly illusory stability. But Moscow also serves notice in many ways that, if necessary, it will proceed without detente, to vie with the U.S. at all points of contact, with what it considers tried and true methods.

SOVIET VIEWS ON A NEW ARMS RACE

Moscow makes it clear that what happens in the military field is central to everything else. It often says it views the growing military might of the Soviet Union as the indispensable underpinning of the favorable correlation of world forces. Moscow categorically insists that the USSR can and will do whatever is necessary to prevent alteration to the present military balance between the U.S. and USSR. It professes confidence that it can succeed.

One basis of its confidence is the time factor. Thus it was recently said, "No matter how hard the American civilian and military leaders try to frighten the Soviet Union with their 'well orchestrated' strategy of counterforce superiority, the United States does not possess this kind of superiority because the majority of the systems on which (the strategy) depends will not be ready for use until the second half of the 1980s."

Beyond this, top Soviet leaders in foreign policy/ security areas insist that the USSR is determined to keep moving in every branch of military activity. At the February, 1981, Party Congress, Brezhnev repeated the standard Soviet line that the USSR does not seek superiority, but " . . . neither will we allow such a superiority to be established over us." Speaking in Kiel on May 9, he added, " . . . if we are compelled, we will find a quick and effective response to any challenge by belligerent imperialism."

In its repeated assertion that it will keep pace in any arms contest with the U.S. — no matter what the scope or duration — Moscow insists that the Soviet economy can stand up to any requirement. Indeed, it claims that it can outlast the U.S. economy in a no-holds-barred arms race. Despite these claims, current *performance* of the Soviet economy indicates that realistic possibilities may be very different. The USSR's own published data on the first six months of its Eleventh Five-Year Plan showed that the USSR is now increasingly beset by serious economic strains.

This data followed an earlier Ministry of Defense monograph by military economist A. I. Pozharov, *The Economic Bases of the Defense Might of a Socialist State*. This monograph, billed as "intended for officers, generals, and other studies of military economics", warns that excessive military allocations will indeed undermine the foundations of Soviet military power. Pozharov argues that excessive military spending " . . . could decelerate the development of the very bases of military power — the (general Soviet) economy — and therewith inflict irreparable damage on the defense capability." He contends that excessive military manpower has two adverse effects. It decreases the pool of workers available to the civilian sector, limiting the potential for general economic development. Also, partly as a result of this drain of manpower, it decreases the workforce and production equipment available to meet warfighting production requirements. In sum, Pozharov takes issue with what he calls " . . . the popular expression that three things are necessary for war: money, money, and even more money."

Together, this monograph and the recent Five-Year Plan data raise serious doubts about the official optimism of Soviet leaders on their ability to meet any new challenge the U.S. may offer in the arms competition field. Pozharov's work and actual Soviet economic performance indicate a radically different situation than that depicted by Soviet leaders. Pozharov indicates that there are limits to Soviet defense expenditures which, if exceeded, would result in serious damage to the Soviet economy on which the USSR's defense capability depends. Other works by military economists imply that these limits have already been reached.

For the past several decades, then, Soviet leaders have been able to maintain a steady growth in military spending as a result of the high, albeit declining, growth rate of their economy as a whole. It was aided in this by the failure of its opponents, chiefly the U.S., to present it with special challenges in either the armaments or political fields. Thus it has been able to set its own priorities in accord with its own objectives.

Now, however, the policies of the Reagan Administration—particularly with respect to military buildup—confront Moscow with a serious problem of resource allocation. Traditionally, resource allocation problems have provoked intense policy disputes with the Soviet leaderships. This new problem is surely no exception; judging from the documents being emitted from the Soviet Union, these disputes are already underway.

Soviet sources clarify that the poor prospects for

their industrial growth arise from the same complex of factors that have plagued a succession of Soviet Five-Year Plans: growing constraints on manpower, capital, and energy; built-in obstacles to technological innovations; rising difficulties in acquiring and distributing raw materials; inflexible, inefficient planning and management; and, of course, the rising burden of defense.

Another aspect of a thoroughgoing U.S. arms challenge to the USSR raises a serious question about official Soviet optimism: i.e., the possibility that the U.S. will make full use of its industrial and technological superiority. It is important to note that, despite the painstaking Soviet effort to deride the possibility that any meaningful breakthrough can be made in weapons technology, Soviet authorities give indisputable evidence that Moscow is, in fact, deeply fearful of precisely that.

A foremost Soviet student of the impact of science and technology on military affairs wrote in *Communist of the Armed Forces* in September, 1974:

Inasmuch as there are no limits to understanding natural laws, so there can be no limit to the application of these laws in technological designs. From this point of view, any, the most terrible, weapon cannot be called absolute since in its stead can come a still more powerful one. . . .

In 1976, a major article in the Soviet *International Affairs* argued that military hardware " . . . must either keep pace with the rapid development of the scientific and technical revolution, constantly absorbing its latest achievements in every field . . . or risk being converted into a pile of rubbish."

Very likely, the Kremlin views the situation as involving something far more than simply outstripping or keeping up with the U.S. by doing more of what each is now doing. In particular, Moscow now apparently sees alarming possibilities for the USSR, such as:

- Being forced into lines of effort the Kremlin did not choose, as a result of new types of U.S. military capabilities or, at minimum, having to change its elaborate and complex war-readiness and war-fighting plans.

- Having its industrial and technological resources taxed beyond their capacity.

- Seeing fractured its basic strategic design of keeping the U.S. so boxed-in militarily as to make it irrational for the U.S. to use our nuclear power for anything other than a self-defeating retaliation for a massive first strike against our own territory.

HIGH FRONTIER'S IMPACT ON THE USSR

In the Soviet view, the High Frontier would confirm its worst fear about U.S. military purposes in space. It would view the move as what the Soviets themselves have characterized as a possible "absolute weapon", capable of ensuring the U.S. "invincibility" from missile attacks. The Kremlin would naturally consider our fulfillment of this aim as some years away. Yet at best, knowing the state of technological art involved and the U.S. capability to build on existing technology, the Kremlin would recognize that we could have substantial capabilities in place within two to four years.

And the Kremlin has always looked at space in terms of military use. Brezhnev explained the settled Soviet approach on October 22, 1969: "Our country has at its disposal an extensive space program calculated for many years to come. . . . We are going our own way, and we are proceeding consistently and purposefully." He then spoke glowingly of an entirely new Soviet reach in space: "Orbiting stations will provide a highway into outer space—they can become 'cosmodromes in cosmos', launching ramps for flights to other planets." Unmentioned by Brezhnev but frequently made explicit by other Soviet writers, the "cosmodromes" would give the USSR means to command the near-Earth environment for military purposes, with inestimable consequences for the strategic balance. And the USSR has, over the last decade, strongly pursued the goals set by Brezhnev. It now appears close to the point where the USSR can hope to serve up a possibly shocking space surprise for the entire Western world, if things continue as they are.

Meanwhile, the Soviets have consistently charged that U.S. space activities are directed toward military ends. This has been good propaganda from Moscow's standpoint, serving not only to further its efforts to stamp the U.S. with an indelible threat-to-peace image, but also to blunt the impact of U.S. successes in space and provide a counter to the attention to the military character of their own program. But implicitly reflected in these charges is the genuine Soviet fear that the U.S. might in fact be using its technological prowess to beat the USSR in using space for military purposes. Recently, unlike earlier periods when charges were made about U.S. military space aims in very general terms, Soviet spokesmen have tended to be very specific in detailing actual U.S. plans, programs, and purposes.

The Kremlin manifests deep concern to avoid a weapons contest with the U.S. that would center on high technology to establish an effective military presence in space. The Soviets want theirs to be the only military presence on the High Frontier, at the pace they choose. The last thing the Kremlin wants is a trial of strength or proficiency in this area. We can safely say that in the area of high technology for advances in weaponry, the Kremlin perceives its greatest single vulnerability compared with the U.S. There can be no doubt that the Kremlin would view the High Frontier as just such a trial.

It may well be that the existing Soviet capabilities are adequate for the USSR to mount a High Frontier system of its own. This in itself would not interfere with or affect the U.S. system in any way. The only effect would be to have antimissile systems on both sides, each incapable of harming or affecting the other. In that case, the strategic situation of the USSR will have been profoundly altered. Instead of enjoying a one-way system of deterrence based on an illusory MAD concept, it would have to adjust to a two-way system based on Mutual Assured Survival. Such a change would wreak havoc with strategic elements of the Soviet warfighting edifice that Moscow has built up over the past 20 years.

One possible Soviet reaction that we must consider is that the Kremlin might simply refuse to stand by while the U.S. puts its new High Frontier systems into operation. It might instead take advantage of the "window of opportunity" that stands open as a result of present Soviet military advantages.

It is a hard fact of our lives that a window of opportunity is open to the Kremlin. That is why something like the High Frontier is so obviously necessary if the U.S. is to ensure its long-term survival without increasing subservience to the USSR. The issue now is whether that window looks big enough for the Kremlin to calculate that the USSR can move through it without unacceptable consequences to itself. If so, a significant probability must be allowed that the Kremlin will use the window at any time it is provoked or opposed by the U.S. If Moscow believes that its window of opportunity is too dangerous, then a strong probability must be allowed that the Kremlin would not risk its use even if faced with the certainty of complete success for High Frontier; for High Frontier will not in any way threaten the existence or even the well-being of the USSR. As a senior Soviet diplomat said in the early 1970s: "ABMs don't kill people; ICBMs do." All the High Frontier will do is deny to any power the ability to threaten the existence of the U.S. Given this, all that is known about Soviet leadership indicates that it is highly likely to choose to wait until another, surer day.

The Soviets' view of the High Frontier, then, will be colored by their need to revise old assumptions, old priorities. If backed effectively, a High Frontier decision by the U.S. would strongly impact the Soviets. Soviet leadership would consider High Frontier as offering the best chance—if not the only chance—for realization of President Reagan's stated intention to refurbish U.S. military power to the point necessary for an effective foreign policy. These conclusions are not assumptions or conjectures. They are the product of exhaustive examination of what the Soviets themselves say and do, and of issues that would be raised by a U.S. effort along the line of High Frontier.

On the basis of testimony we have cited from Moscow, in Kremlin eyes a credible U.S. commitment to High Frontier would:

- Confront the USSR with precisely the sort of competition Soviet leaders most fear, and are most anxious to avoid.

- Severely tax, perhaps to the point of disruption, the already strained Soviet technological and industrial resources.

- Seriously threaten the very foundations of the strategic structure the USSR has built at great cost over the last 20 years.

- Undercut the foundation for Kremlin claims that the "correlation of world forces" has irreversibly shifted in favor of the USSR.

- Force a Soviet restructure of doctrinal concepts and strategic designs that Soviets have structured over the years for victory over the West.

With this examination of Soviet views behind us, we can predict with virtual certainty that Moscow would pull out all stops in its vast pipe organ of propaganda in an effort to deter or deflect the U.S. from the High Frontier. Moscow's criticism would bring to bear all of the arguments it now employs against alleged U.S. intentions to inaugurate a new arms race. It would target a lot of attention to the militarization of space issues; it would charge that we were violating the Space Treaty of 1967, the ABM

Treaty of 1972, dozens of UN resolutions, the UN Charter, and so on. The fact that High Frontier systems are nonnuclear would not affect this propaganda. Moscow would disregard unpleasant facts in feeding its propaganda mill and would treat as truth anything it could dream up—as it is already doing with regard to the Space Shuttle.

SUMMARY

From the foregoing, it is evident that High Frontier will receive critical attention from various sources and of several kinds. Well-reasoned, expert criticism is the stuff of which High Frontier was forged, and more of it may provide a broader base for initial systems evaluation when and if U.S. leadership deserts Mutual Assured Destruction in favor of Assured Survival.

We devoted considerable space to expectations of the Soviet view because, while Soviet criticisms are likely to be freely offered, the Soviets have given us ample reason to doubt their fair intentions. The Soviet military presence in space is a fact we take for granted. The American military presence in space is an issue the Soviets brand as saber rattling. When sufficient fair criticism is brought to bear on the High Frontier, it will be obvious that we propose, not a saber, but a shield.

CHAPTER XII
THE WEST EUROPEAN VIEW

Our adoption of the High Frontier strategy would lead to a revival of faith for responsible West Europeans, if our decision were backed by convincing evidence of a U.S. determination and ability to make it work. The effectiveness and reliability of the High Frontier would be viewed as a godsend to their security interests.

From the West European standpoint, deterrence as a viable strategic concept has always depended upon whether it would safeguard Western Europe from Soviet attack. As long as the U.S. had a monopoly or near-monopoly of nuclear power, Europeans took for granted that this would be the case. Both the logic of the importance of Western Europe to the U.S., and our commitment to its defense through NATO confronted the USSR with the apparent certainty of nuclear devastation in case it resorted to overt aggression in Europe.

However, as the USSR developed mounting capabilities to inflict nuclear devastation on the continental U.S., skepticism has grown among Europeans that the U.S. would actually risk its own destruction in their defense. The Soviets have constantly fed this skepticism, contending that the U.S. would be deterred from using its nuclear power not only because it would be destroyed by a nuclear war, but also because the capitalist system as a whole would be destroyed. Moreover, since the late 1950s, the West Europeans have heard a persistent argument from the Soviets, i.e., that whatever the U.S. did, it would not help the Europeans. This Soviet claim asserts that the first and surest result of any nuclear war between the U.S. and USSR would be the total destruction of all European countries allied with the U.S. in that war.

In European eyes, the U.S. movement toward a doctrine of Mutual Assured Destruction appeared to confirm their fears. If, in fact, the U.S. was prepared to accept that any use of nuclear weapons would constitute an act of suicide on its part, then it was simply inconceivable that the U.S. would employ these weapons in any circumstance short of a direct attack on its own territory. Many doubted that we would do so even in *that* extreme, perhaps judging from U.S. failure to implement Civil Defense measures which Europeans themselves consider essential.

In case of a Soviet advance into Western Europe, the U.S. stance appeared dangerously ambiguous to the Europeans. On the one hand, there remained the U.S. commitment to participate in the defense of Europe, and to provide theater nuclear weapons for that purpose, including those capable of reaching key targets in the USSR. On the other hand, some doubted that the U.S. would respond to a Soviet attack by immediate employment of its full arsenal of strategic weapons. This seemed to mean that, at best, the U.S. would put its own territory at risk only after the West European countries had been devastated by theater nuclear exchanges. The worst fear was that the U.S. would simply hold back its strategic nuclear power in the hope of deterring the USSR from nuclear strikes against a North America devoid of any true defense, civil or otherwise. Thus deterrence for the Europeans has come to be a strategy of faith, rather than one of reasonably assured successful defense.

Officially, West European governments have continued to adhere to the MAD strategy because, bluntly, they have had no alternative. However, they increasingly give reasons for doubt that they would act out their role in the U.S.-conceived scenario of theater defense, should the USSR decide to test that scenario. The current attitudes of the peoples of West European countries justify an assumption that the governments would have real difficulties in offering resistance on the order planned, even if they decided to do so. Careful analysis indicates that these attitudes reflect a pervasive popular feeling among West Europeans, and are not simply a display of contrived anti-Americanism on the part of small groups of extremists.

As the current antinuclear leftist campaign in Western Europe grows steadily stronger, the political prospects for NATO Theater Nuclear Force modernization become less favorable. The incremental approach to improving the situation with additional nuclear weapons, as dictated by the U.S. strategic doctrine of deterrence, now seems to be counterproductive. A new answer to NATO military problems is becoming absolutely necessary.

A new U.S. doctrine which shifts emphasis from an exclusively offensive stance to one in which offense is balanced by genuine defense would eventually appeal to our European allies. It would make the Soviet nuclear weapon threat less effective, giving Europeans reassurance of survival rather than the certain-death scenario of Mutual Assured Destruction. It would not, of course, turn the antinuclear, anti-U.S. element completely around. That group will quickly regear—perhaps lubricated with arguments made in Moscow—to blast the new strategy as provocative and likely to increase the probability of nuclear war.

A new, genuinely defensive strategy, carefully conceived and presented as one designed and phased to gradually strengthen deterrence, should command the support of responsible European political leaders, defense-knowledgeable elites, and many opinion leaders in Western Europe. This is particularly true if the new strategy can take the world out of the shadow of nuclear incineration, and yield important commercial benefits as well.

High Frontier would thus profoundly alter the strategic situation of Europeans for the better. Its formal announcement—or even its quiet inauguration—would not in the short term quell the popular turmoil in Europe. Indeed, the immediate effect might be to increase the turmoil because of fears that the USSR might be provoked into some sort of action, the brunt of which might fall first or foremost in Europe. Further, there would be strong reactions by many to the danger of extending the weapons contest into space. Some government leaders might urge delays in the hope of not rocking the boat at this particular juncture. A hard core of pro-Soviet elements intermixed with other Europeans—their equivalents of U.S. diehard believers in the usefulness of negotiated arms control—would remain hostile to the High Frontier.

But for most Western Europeans, a demonstration that the U.S. was determined in this new chosen course would lead to the following decisive changes in European attitudes in a relatively short time:

- There would be a realization that the U.S. was beginning to break out of the paralytic bonds imposed by the concept of Mutual Assured Destruction.

- Even more decisive would be the realization that the High Frontier would provide protection for Europe from Soviet-launched strategic ballistic missiles. The current absence of such protection, and the resultant forced exclusive reliance on the deterrent effect of threatened retaliation, have been the chief sources of the great nightmare with which Europeans have had to live for years—since the USSR began its deployment of a multitude of medium-range missiles capable of reaching all points of Europe and adjacent territories. For the Western Europeans, this would mean a strategic turnaround of momentous proportions.

- Finally, there would be a restoration of the badly shaken European confidence in the U.S. ability to resolve to actually use its power to preserve the Free World.

CHAPTER XIII
TREATY CONSIDERATIONS

The High Frontier programs cannot be implemented without an impact on arms-control negotiations past, present, and future. At the core of High Frontier is a fundamental change from MAD, and we cannot make this strategic change without also rethinking our approach to arms control.

U.S. arms-control efforts to date have been rooted in two basic percepts of MAD: that stability depends on a balance of terror sustained by a negotiated equality (or parity, equivalence, sufficiency, etc.) in *punitive* nuclear weapons; and the inescapable corollary of that doctrine, that strategic *defensive* weaponry is destabilizing and provocative.

The present MAD-based approach to arms-control negotiations was articulated clearly in a Pentagon news conference on May 18, 1967, by Secretary of Defense Robert S. McNamara. He stated:

We think it is in our interest, and theirs, to limit the deployment of defensive weapons, and we're quite prepared to discuss possible limitations in the deployment of offensive strategic nuclear weapons as well.

Mr. McNamara calculated that the Soviets would be unlikely to embrace "balance of terror" as a basis of arms-control negotiations as long as the U.S. maintained a superior position in strategic nuclear offensive power. He asserted that strategic stability and the conditions for effective SALT negotiations would be improved if the Soviets were allowed to increase their nuclear attack capabilities to a level where they would be certain of inflicting intolerable destruction on the U.S. in a retaliatory strike. He moved forcefully within the Department of Defense to defeat strategic programs, defensive or offensive, which could thwart the achievement of this presumably desirable balance of terror.

THE INFLUENCE OF MAD ON SALT

The results of SALT negotiations, thus far, quite clearly demonstrate the real effects of these MAD precepts. The only *treaty* resulting from SALT is the ABM Treaty which is designed to outlaw strategic defense—at least defense against ballistic missiles, which are the most threatening of offensive systems. On the other hand, our attempts to negotiate limits on offensive systems (the Interim Agreement of 1972, the Vladivostok Accords, and SALT II) resulted in the *escalation* of offensive nuclear weapons! Americans were urged to accept this "progress" in SALT on the basis of the MAD theory—that nuclear war would be so apocalyptically destructive that its deterrence is independent of the numbers of weapons involved.

The U.S. has entered negotiations on offensive systems with attempts to fix limits consistent with MAD; i.e., at or below existing U.S. weapon inventories, hoping to avoid any increase. The Soviets, in contrast, enter these negotiations determined to fix levels high enough to accommodate an entirely different strategy. That Soviet strategy insists that nuclear war would destroy capitalist nations, but that the socialist camp —despite widespread destruction—would emerge triumphant, in part because of their very large advantage in active defenses and Civil Defense preparations. Soviet SALT negotiators insist that nuclear weapon levels be high enough to encompass their ongoing weapons programs, which are designed to support that strategy.

Invariably, both sides are accommodated. The Soviets are allowed to pursue the war-winning capabilities consistent with their doctrines of winnable war, while the U.S. is permitted to add to its retaliatory-only capabilities consistent with MAD doctrine. The inevitable effect of negotiations based on these two fundamentally divergent strategies has been an intolerable growth of Soviet nuclear first-strike capabilities with effective strategic defenses, coupled with a dangerous weakening of the U.S. deterrent with virtually *no* defense at all. Western arms-control advocates have been unwilling to accept an obvious reality: the Soviet Union rejects, in word and deed, the MAD doctrine which underpins Western devotion to the SALT process.

From the very inception of the MAD theory, the Soviets have branded it as "bourgeois naivete." They have not chosen to leave their homeland defenseless against U.S. nuclear retaliation, as MAD theory demands. Instead, they have poured more resources into strategic defenses—both active and civil—than the U.S. has invested in its entire deterrent force. They have created offensive systems obviously designed to destroy as much as possible of the U.S. retaliatory force in a first strike.

This incontrovertible evidence of the Soviet strategic perspective, and its incompatibility with the U.S. approach to SALT negotiations, is swept aside by many arms-control advocates by a far-fetched assumption that there are Western-style "hawk" and "dove" factions competing for control in the Kremlin. According to this assumption, top Soviet "civilians" really do accept MAD theory, but are opposed by a powerful group of Soviet "militarists" who insist that nuclear war is not only thinkable but winnable.

To accept this view of Soviet leadership, one must somehow believe that the omnipotent Communist Party, headed by Yuri Andropov, cannot control its comrades in the Red Army and indeed must reluctantly imperil the entire economy of the USSR to meet the demands of a Russian military/industrial complex. If one can bring himself to believe this, he can then argue—as many arms-control advocates do—that the U.S. should accommodate intransigent Soviet positions in order to strengthen the hand of Kremlin "doves" in their difficult task of restraining their military "hawks". It would also follow that the key alliances in the arms-control process are, on the one hand, between peaceloving adherents of MAD in the U.S. and USSR and, on the other hand, between the militarists in the Pentagon and their counterparts in the Soviet Ministry of Defense.

As long as MAD remains the basis for the U.S. approach to arms control, the SALT process will continue to undermine the security of the Free World. So long as SALT is pursued by U.S. negotiators attempting to establish and maintain a balance of terror, and Soviet negotiators determined to establish and maintain strategic nuclear dominance, no SALT agreement will ever be ratified by the United States Senate.

Further, the longer MAD theory and arms-control advocacy remain interlocking concepts among Western intellectuals, the more contrived will be the excuses provided for Soviet behavior—whether in the SALT process or elsewhere.

CURRENT TREATIES

The High Frontier strategy of Assured Survival can be adopted and pursued without regard for further arms-control agreements with the Soviets. In fact, one of the salient advantages of High Frontier is that it provides security to the West quite independently of any trust or distrust we may feel toward Soviet leaders. The usefulness of High Frontier's spaceborne strategic defenses are not affected by Soviet compliance with past arms-control agreements. This important advantage should not be affected by any future arms-control agreements.

This is not to say that the High Frontier strategy excludes all consideration of arms control. In fact, the High Frontier approach has received support from an unexpected quarter. Anders Boserup, a Danish activist in the international disarmament movement, stated in the *Bulletin of the Atomic Scientists*, December, 1981:

> . . . the adoption by states of a defense approach to security need not lead to an arms race. On the contrary, it can lead to disarmament, and is probably the only viable approach to it.

One aspect of survivability *not* treated in the High Frontier study is the advantage to be gained by applying stealth technology and decoys to the GBMD concept. It is the opinion of some of our team that for relatively small exertion on our part with stealth and decoys, the Soviet problem of trying to counter GBMD can be enormously complicated.

Most important in the consideration of survivability is to bear in mind the strategic reasoning behind the deployment of a spaceborne defense in the first place. Its primary strategic purpose is *to deter a Soviet first strike*. The purpose is *not* to survive, not to prevent a Soviet retaliatory strike, and not to eliminate totally the Soviet missile threat. When this basic fact is borne in mind, it is clear that even *should* the Soviets devise a means to knock a hole in the spaceborne defensive satellite system through which to launch an ICBM attack on the United States, they could not expect a first strike to succeed. They would never believe that after an assault on our defensive systems, our offensive weapons would await destruction on the ground. Thus, even if one postulates an effective Soviet attack mode against GBMD, it would not destroy the *basic strategic* reason for deploying it.

It should also be noted that some of the postulated Soviet counters to GBMD involve massive investment and long lead times. For instance, Aerospace Company has suggested that the Soviets might mass their ICBMs in a small piece of geography in order to "bore through" the defense. That is a physical possibility, of course, but it is fraught with adverse strategic considerations for the Soviet planner, would cost enormously, and would take a decade to accomplish even partially. And—again—the Soviet first-strike threat would be negated for the foreseeable future.

The big problem with High Frontier isn't technology or costs, it's the effects of it on arms control, especially the ABM Treaty.

We expected a good deal of concern about the effects of High Frontier on the ABM Treaty and devoted considerable space to the issue in our study. But we have found as we have travelled the country that it is a non-issue with the public and the press. Curiously, we hear most about it from the Army and the Air Force. Secondly, we have found that the public is highly supportive of defense against Soviet ICBMs. A recent poll showed that two out of three Americans believe we already *have* a defense against ICBMs; 80 percent said they want such a defense and that cost would *not* be a primary factor in their support of such a system.

It may well be, as Mr. Boserup says, that agreements will become possible only when defensive weapons are reemphasized. Consider this. If the United States adopted the High Frontier recommendations, our arms-control negotiators could make a proposal to the Soviets along these lines:

The most destabilizing of weapon systems on both sides are long-range ballistic missiles. Only by using these weapons could either side ever hope to carry out a successful first strike against the other's homeland. We have declared that we have no intention of delivering a first strike against the USSR, but perhaps you do not consider that an adequate guarantee on which to base the security of the USSR and the Socialist Bloc. If nothing else, your heavy expenditures on strategic defenses and civil defense so indicate. You have also stated that you have no intention of launching a first strike against the United States, but we can also reasonably doubt that such a declaration is adequate to insure the security of our country and our allies.

We are going to deploy a purely defensive, spaceborne defense to ensure that we cannot be subjected to a nuclear first strike by any nation or combination of nations. It will be a nonnuclear system which cannot possibly be used for attack on any person on earth. We invite you to deploy a similar system to prevent any nuclear first strike against the USSR. We believe that our current treaty arrangements should be amended to make it quite clear that these bilateral steps toward a stable strategic situation are mutually acceptable. We are willing to agree to measures which will ensure that the systems deployed are unmistakeably defensive and nonnuclear.

If we can both look forward to the day when neither side must constantly guard against a first strike on our retaliatory forces and our nations as a whole, then surely we can reduce the total numbers of nuclear weapons on both sides, as President Reagan has proposed.

This kind of arms-control offer has some chance of success. It is not based, as U.S. arms-control efforts thus far have been, on the hope of creating more balance in the balance of terror, required by the Mutual Assured Destruction theory, but on creating a situation of Mutual Assured Survival. Total nuclear weapons stockpiles would be effectively reduced by applying technology which makes the accumulation of offensive nuclear weapons less important to both parties.

True, the Soviets may very well reject the entire proposal. But in doing so, they will declare themselves unwilling to accept any solution which denies them a first-strike capability against the United States. Thus we must proceed with High Frontier whether or not they agree to work for Mutual Assured Survival.

If we choose to reemphasize strategic defenses, how do we know that the Soviets won't quickly deploy a lot of ABMs?

Ironically perhaps, the employment of effective spaceborne defenses will accomplish, through unilateral U.S. action, that same result which the disarmers have so fruitlessly pursued for over a decade and a half of SALT talks—the checking of the growth of nuclear offensive weapon inventories on both sides. Effective strategic defenses can negate the paramount importance of the nuclear ballistic missile in the strategic equation, and can eliminate the imperative on both sides to have more weapons with even greater destructive power. The U.S./USSR competition would be shifted from a numerical contest in nuclear offense to a technical contest in defensive systems in space where technologies show great promise—including tremendous *peaceful* promise.

Even if nuclear weapons come to play a role in the defensive competition in space, the threat of their use, hundreds of miles above the Earth, would certainly be preferable to the threat they now pose in the form of warheads aimed at terrestrial targets.

Despite these advantages for the world of peace and security from nuclear devastation, the adoption of the High Frontier strategy will require a fundamental change in our approach to arms-control negotiations; and this is certain to engender controversy. MAD theories will not die easily, in or out of government. There is no bias among bureaucrats stronger than that bias toward the devout support of positions taken in the past. A myriad of interlocking policies and positions taken in the State Department, DOD, and the Arms Control and Disarmament Agency would require drastic revision if the U.S. approach to arms control is to be based on a search for Assured Survival, rather than for a perpetual balance of terror.

Of immediate concern in the area of arms control are those treaties which address the uses of space and strategic defensive systems—the Outer Space Treaty, negotiated under UN aegis in 1967, and the ABM Treaty between the U.S. and USSR signed in May, 1972. The preamble to the Outer Space Treaty refers to three UN General Assembly resolutions which cover "Legal Principles Governing the Activities of States in the Exploration and Use of Outer Space", a call upon states " . . . to refrain from placing in orbit around the Earth any objects carrying nuclear weapons or any other kinds of weapons of mass destruction . . . " and the condemnation of propaganda, which would " . . . provoke or encourage any threat to peace. . . . " The Outer Space Treaty is essentially consistent with these UN General Assembly resolutions.

KEY POINTS OF OUTER SPACE TREATY

- The exploration and use of outer space, including the Moon and other celestial bodies, shall be carried out for the benefit and in the interest of all countries.

- States Parties to the Treaty shall carry on activities in the exploration and use of outer space . . . in accordance with international law.

- States Parties to the treaty undertake not to place in orbit around the Earth any objects carrying nuclear weapons or any other kinds of weapons of mass destruction, install such weapons on celestial bodies, or station such weapons in outer space in any other manner.

- The Moon and other celestial bodies shall be used by all States Parties to the Treaty exclusively for peaceful purposes.

- States Parties to the Treaty shall bear international responsibility for national activities in outer space.

- If a State Party to the Treaty has reason to believe that an activity or experiment planned by it or its nationals in outer space . . . would cause potentially harmful interference with activities of other States Parties in the peaceful exploration and use of outer space . . . it shall undertake . . . consultations before proceeding with any such activity or experiment.

The preamble to the ABM Treaty ties defensive and offensive weapons limitations together with such lan-

guage as: " . . . the premise that nuclear war would
have devastating consequences for all mankind";
" . . . effective measures to limit antiballistic missile
systems would be a substantial factor in curbing the
race in strategic offensive arms and would lead to a
decrease in the risk of outbreak of war involving nuclear weapons"; " . . . measures with respect to the
limitation of strategic offensive arms would contribute to the creation of more favorable conditions for
further negotiations on limiting strategic arms". It
declares as an intention of the ABM Treaty: " . . . to
achieve at the earliest possible date the cessation of
the nuclear arms race . . . " and " . . . general and complete disarmament. . . . "

KEY POINTS OF ABM TREATY

Each Party undertakes not to deploy ABM systems
for a defense of the territory of its country and not
to provide a base for such a defense, and not to
deploy ABM systems for defense of an individual
region except as provided for in Article III of this
Treaty.

The Treaty describes its purpose as to counter strategic ballistic missiles in flight trajectory. It clearly
identifies an ABM system, for purposes of the treaty,
as (1) ABM interceptor missiles, (2) ABM launchers,
and (3) ABM radars. It limits deployment of one ABM
system around a " . . . Party's national capital . . . " and
another ABM system to deploy around ICBM silo
launchers.

Each Party to the treaty agrees " . . . not to develop,

test, or deploy ABM systems or components which are sea-based, air-based, space-based, or mobile land-based". The treaty limits the launching of one ABM interceptor missile at a time for launchers.

Each Party agrees " . . . not to deploy in the future radars for early warning of strategic ballistic missile attack except at locations along the periphery of its national territory and oriented outward."

The 1974 summit protocol further limits the number of ABM sites to one in each country. (The U.S. has none, having abandoned the Grand Forks site. The Soviets defend Moscow with their one site.)

The ABM Treaty narrowly defines the ABM system as interceptor missiles, launchers, and radars. This "tight" definition was addressed by the U.S. and Soviet heads of delegation on May 26, 1972, the same date as the original Treaty signing, in "Agreed Statements".

In order to insure fulfillment of the obligation not to deploy ABM systems and their components except as provided in Article III of the Treaty, the Parties agree that in the event ABM systems based on other physical principles and including components capable of substituting for ABM interceptor missiles, ABM launchers, or ABM radars as created in the future, specific limitations on such systems and their components would be subject to disagreement in accordance with Article XIII and agreement in accordance with Article XIV of the Treaty.

Articles XIII and XIV refer to meetings of a Standing Commission to consider measures bearing on the

Treaty as well as establishing the consideration of amendments proposed by either Party. Parties will conduct a review of the Treaty every five years.

HIGH FRONTIER IMPACT ON TREATIES

With regard to the UN treaty on outer space, nothing in the High Frontier concept contradicts its language. The prohibition against weapons of mass destruction in orbit is not violated by any of the High Frontier military programs involved, and the nonmilitary programs can be fairly depicted as beneficial for all countries.

Even so, the U.S. government would have to prepare for a polemical buffeting by the Soviet Bloc and its Third World clients for engaging in "space imperialism". The linchpin for such a propaganda assault has already been set by the Soviets in their proposed new UN treaty outlawing *all* space weaponry. (Sad to say, this proposition is now supported by some members of Congress!)

A more serious problem for High Frontier is presented by the ABM Treaty. As the only real treaty to emerge from the SALT process, it is of great symbolic value to arms-control advocates. It also represents the legalistic refuge for adherents of the MAD doctrine. Finally, it was negotiated, ratified, and applauded by many influential figures from many quarters of the U.S. body politic.

High Frontier represents a direct refutation of the philosophical basis underlying the ABM Treaty. The defensive systems advocated by High Frontier do not necessarily conflict with the specific provisions of the treaty, but they can and will be construed as conflicting with both the spirit and the letter of it.

There are three basic legitimate answers to real or alleged conflict between High Frontier and the ABM Treaty: abrogate the treaty, assert that we are in compliance with it, or amend it.

ABROGATION

The ABM Treaty provides for withdrawal by either party in the event that its "supreme interests" are jeopardized. The U.S. Senate was assured in 1972, prior to ratification, that failure to achieve progress in offensive strategic weapons limitation agreements would be grounds for U.S. withdrawal. Certainly the case can be made that SALT negotiations have failed to check the unprecedented growth of Soviet nuclear offensive power, and that this jeopardizes U.S. supreme interests. Add to this the strong evidence of Soviet violations of this treaty, and the case for abrogation is clear.

ASSERTION OF COMPLIANCE

The definitions of what constitutes an ABM system within the context of this treaty are rather rigid. The spaceborne ballistic missile defense systems involved in the High Frontier concept do not fall within those definitions, and can be fairly described as "ABM systems based on other physical principles". Limitations on such systems become the subject of discussion between the signatories. Such discussion can be initiated without hindering U.S. actions to acquire such systems, and can be directed toward permitting, not denying, their deployment.

The case can also be made that certain point defense options in the High Frontier layered defense concept also fall outside treaty definitions of ABM. In any case, at least 100 U.S. ICBM silos could be protected against a first strike without violation of the treaty.

AMENDMENT

The 1972 ABM Treaty provides for review and amendment every five years. The 1977 review was only perfunctory. In subsequent reviews, the U.S. negotiating team should propose amendments to permit unfettered U.S. acquisition of defensive systems if the options of abrogation or asserted compliance are rejected, or appear inadequate to support the High Frontier efforts.

Best-Selling Science Fiction from TOR

Best-Selling Science Fiction from TOR

☐	48-549-2	**Paradise** Dan Henderson	$2.95
☐	48-531-X	**The Taking of Satcon Station** Barney Cohen and Jim Baen	$2.95
☐	48-555-7	**There Will Be War** J.E. Pournelle	$2.95
☐	48-566-2	**The McAndrew Chronicles** Charles Sheffield	$2.95
☐	48-567-0	**The Varkaus Conspiracy** John Dalmas	$2.95
☐	48-526-3	**The Swordswoman** Jessica Amanda Salmonson	$2.75
☐	48-584-0	**Cestus Dei** John Maddox Roberts	$2.75

Fred Saberhagen

☐	48-501-8	The Water of Thought	$2.50
☐	48-564-6	Earth Descended	$2.95
☐	48-520-4	The Berserker Wars	$2.95
☐	48-536-0	Dominion	$2.95
☐	48-539-5	Coils *with Roger Zelazny*	$2.95
☐	48-560-3	**The First Book of Swords** *March 83*	$6.95